PHYSICAL CHEMISTRY

Statistical Mechanics

PHYSICAL CHEMISTRY
Statistical Mechanics

Horia Metiu
University of California
Santa Barbara

New York • London

Vice President	Denise Schanck
Senior Editor	Robert L. Rogers
Assistant Editor	Summers Scholl
Senior Publisher	Jackie Harbor
Production Editor	Simon Hill
Copyeditor	Ruth Callan
Cover Designer	Joan Greenfield
Indexer & Typesetter	Keyword Group
Printer	RR Donnelley

Cover image courtesy of Jeffry Madura

ISBN 0 8153 4085 0

Library of Congress Cataloging-in-Publication Data

Metiu, Horia, 1940-
Physical chemistry : statistical mechanics / Horia Metiu.
p. cm.
ISBN 0-8153-4085-0 (acid-free paper)
1. Statistical mechanics–Textbooks. 2. Chemistry, Physical and theoretical–Problems, exercises, etc. I. Title.

QC174.8.M45 2006
530.13–dc22

2005032154

Published in 2006 by Taylor & Francis Group, LLC,
270 Madison Avenue, New York, NY 10016, USA and
4 Park Square, Milton Park, Abingdon, Oxon, OX14 4RN, UK.

Printed in the United States of America on acid-free paper.

10 9 8 7 6 5 4 3 2 1

CONTENTS

PREFACE

Most people write a textbook because they have not found one that presents the appropriate material in the proper way. I am no exception to this rule and here is why. When I started teaching Physical Chemistry, I was disappointed by the books available to me. We implement physical chemistry, in our professional work, by using extensive computations. Until very recently these were too laborious for classroom work. As a result, textbook writers developed what I call "pedagogical physics" or "physics with abacus": a version of physical chemistry watered down to allow doing homework with a calculator. In such books, the ideal gas law reigns supreme; all mixtures are ideal; the temperature dependence of heat capacities, heats of reaction, and entropies are ignored; the Clapeyron equation is mutilated to make it easy to integrate; the calculations of equilibrium composition are confined to the simplest reactions, which would not lead to high-order algebraic equations; essential topics, such as competing reactions, have to be avoided.

After making such simplifications, it makes no sense to compare theory with experiments. Unless a sanitized version of reality is chosen (low pressure and high temperature, small temperature ranges, very simple reactions, etc.), the calculations disagree with the measurements, giving physical chemistry a bad name. As a result, the textbooks fail to show how the theory being taught adds value to the experimental work or to technology.

I believe that we can change this situation and teach a more realistic physical chemistry, than we currently do, by using software that allows the student to overcome the fear of mathematics and the tedium of the computations. I have in mind symbolic manipulation languages, such as **Mathematica®** or **Mathcad®**, which make it possible to easily perform the calculus manipulations that physical chemistry requires, and to quickly write programs that perform extensive numerical calculations on a personal computer. Since most students own a computer these days, using one in the class room no longer requires the existence of a computer laboratory in the department.

Since such computing is possible, a new textbook is needed in which the simplifications are removed (this requires rewriting much of the theory), realistic examples are used (which requires writing programs), and extensive comparisons with experiments are provided (which requires searching the library for data). Moreover, the topics omitted, because they were mathematically or computationally too complicated, need to be reinserted. This book, together with volumes on kinetics, quantum mechanics, and thermodynamics, is an attempt to provide such a textbook.

I developed these volumes while teaching physical chemistry at the University of California at Santa Barbara. It is perhaps useful to share our experience and explain how we have used computing in the classroom. When we started (about six years ago) very few students owned a computer and not a single chemistry student in the class knew how to use a computer to perform scientific calculations. This was rather appalling: it is, in my opinion, unacceptable to give a chemistry degree to people who are unable to perform calculations that require more than a hand-held calculator. It also posed severe constraints on what we could do. My colleague Alec Wodtke decided to help and he created a new "laboratory" course in which he was going to teach the students how to program with **Mathematica**. At the time Alec made this decision, he did not know **Mathematica**, but was eager to learn it for personal use. In about two weeks he became a decent **Mathematica** programmer and was a "master" by the time he finished teaching the class. Learning **Mathematica** can be a rather pleasant and entertaining learning experience. My task was to write new lecture notes and the **Mathematica** programs accompanying them, and to teach the physical chemistry class.

Since we did not have enough computers for all the students enrolled in the course, we created a "**Mathematica** track" and offered it to a group of twenty to thirty students. This caused additional complications: I had to write lecture notes for all students, regardless whether they used a computer or not. In addition, I had to give two sets of homework. This additional labor was a great stroke of luck, because it prevented me from writing a book of physical chemistry dependent upon **Mathematica**. This textbook can be used to teach physical chemistry without a computer; but the experience is enriched substantially for those students who do learn how to read and write **Mathematica** or **Mathcad** programs.

The lecture notes and the **Mathematica** programs were posted on the web and were used in the classroom (no textbook was recommended), for three years in a row. A lack of classroom space made it very difficult to schedule separate lectures to teach programming. Because of this, our students started learning **Mathematica** and physical chemistry at the same time. The tutorial in **Mathematica** was structured so that in a few weeks the students were able to do the homework in the physical chemistry course. The **Mathematica** Workbooks that I posted on the web provided a "programming template" that the students could follow. After two weeks of Alec's tutoring, the students managed to use **Mathematica** to do their homework. While this arrangement worked for us, a more rational approach would be to teach programming in advance of physical chemistry. My preference would be to start immediately after the students finished learning calculus, before they take physics.

The choice of **Mathematica** was not based on "market research." I have used **Mathematica** almost daily since it appeared and I was sure that it would do the job well. As the lecture notes turned into a book, my editor, Bob Rogers, wisely decided to expand our audience and recruited Professor Jeffry D. Madura (Duquesne University) to produce a **Mathcad** version of my **Mathematica** programs. His programs and mine are included in the CD-ROM included with this book. Whether one uses **Mathematica** or **Mathcad** depends mostly on whether the campus has a license for one or the other, and on the personal history of the instructor.

Changing the tools we use, changes the work we do. Comparison with experiments is now possible and it has become an integral part of the course. After all, we teach this material mainly for future use in practice, even though at times we teach to provide amusement and illumination.

Increasing the level of the material presented in the class entails some dangers. The students can easily become lost or frustrated if the pace is too fast or key details are skipped. Because of this, I paid very close attention to clarity. I have tried to determine what an average student can understand, by working closely with the students during office hours. I was also lucky that Celia Wrathall decided that typing the lecture notes provided her with an opportunity to learn physical chemistry. She peppered me with questions, admonitions, and suggestions regarding the clarity of the material and of the phrasing of the text. No paragraph in the book survived if Celia did not understand it.

Making the material clearer is usually a matter of dosage. I tried to outline the line of thought at the beginning, and to break it up into smaller, logically connected parts. I did not hesitate, as an argument developed, to review how far we had gotten and where we were going. I hope that a better understanding is a valuable compensation for increasing the length of the text.

Physical chemistry is where the undergraduate science and engineering student encounters the rigorous, quantitative methods of science. We teach certain material, but we also teach the method of using physics and mathematics carefully and precisely for solving physics problems of interest to chemistry. With this in mind, I avoided sloppy arguments and "linguistic physics," the habit of covering up ignorance with vague nomenclature. If a limitation in the background of the student made it impossible to prove or explain correctly a statement, I said so. I did not try to create the illusion of understanding by using fuzzy language.

One of the most difficult duties of a textbook writer is deciding what material to teach. Severe time constraints make this task harder. In the 1930s physical chemistry was taught for a year and the course covered only thermodynamics and kinetics. Since then, we have added quantum mechanics and statistical mechanics, without adding a single hour of lecture. This puts a tremendous pressure on us to condense, eliminate, and consolidate the content. I have chosen the material to meet the practical needs of the chemist and engineer. In this book the most important chapters are those that explain the meaning of entropy and its connection to information, those on chemical equilibrium and those on transition state theory. Few people will quarrel with the statement that rate constants and equilibrium constants are the most important quantities in chemistry. In this book they received unusually detailed attention.

We teach statistical mechanics and kinetics in one quarter, which makes it difficult to cover all the material in the book equally. I did not try to do that out of fear that by being taught too much the students will learn too little. It is said that Victor Weisskopf once conversed with a young assistant professor who was excited by how much material he managed to present to his class. Weisskopf remarked: "you young people like to cover a lot; I like to uncover a little." It is better to teach less material well, than to go too fast through a lot of material. After all, the purpose of this course is to introduce the students to statistical mechanics in a way that allows them to read a more advanced book when they need to. To accomplish this, one must explain clearly the principles and the manner in which they are used; there is no need to cover all possible applications.

A possibility, which I tend to follow, is to accept that the climax of statistical mechanics, at this level, is its application to the study of chemical kinetics and of chemical equilibrium. To get there I spend little classroom time on the chapters containing comparisons with the experiment, since this material is best left for individual study and home work. I only try to make clear the line of thought and to present in detail the more subtle points.

Many recent textbooks give long lists of references at the end of every chapter. I have tried to avoid this, since long lists increase the entropy of information, not its quality. I give a few general texts at the end of the book, which can be consulted with benefit in case of confusion or offer a good start for further study.

Finally, it is a tradition in preface writing to thank the people who helped the author along the way. I want to thank all the students who came to office hours, since they taught me what the average student knows and what he or she can do. Various teaching assistants have also helped me stay anchored in reality. Professor Michelle Francl (Bryn Mawr) has been a priceless reviewer. Her comments did much to improve the book and her wit made me chuckle often. I am also grateful to Professor Hannes Jónsson (University of Washington), Professor Dmitrii Makarov (University of Texas), and Professor Steve Buratto (University of California at Santa Barbara) for useful advice and encouragement. Professor Flemming Hansen (Technical University of Denmark) also provided helpful comments. My editor Bob Rogers believed in this book and supported it and it has been a pleasure to work with him. Summers Scholl at Taylor & Francis provided useful comments and kept the project on track. My assistant Celia Wrathall typed and read the text, re-derived and tested all equations, asked many questions that helped clarify the text, and was an invaluable partner throughout the writing of the book. I am indebted to her for her intelligence and hard work. This book could not have been written without the love, the support, the humor, and the patience of my wife Jane.

HOW TO USE THE WORKBOOKS, EXERCISES, AND PROBLEMS

Workbooks on CD-ROM

A special feature of this book is the use of symbolic manipulation programs to perform some of the derivations and the calculations that show how to use the theory. These **Mathematica** and **Mathcad** programs are organized into Workbooks, which collect all the calculations related to a given chapter. All chapters that have a Workbook associated with them have an icon placed after their title. For easy reference, these icons also appear in the margin of the textbook page wherever a specific Workbook is first used and should be consulted.

These computer programs are auxiliary materials. Each chapter in the book explains fully how the calculation is done and gives the necessary intermediate steps and results. The student can follow them without consulting the Workbook. In addition, every program in the Workbooks contains a summary of the theory used in the calculation and can be understood by a student who has not learned programming. To help students who do not have much programming experience the programs are as simple as possible. If a more sophisticated syntax is necessary, its meaning is carefully explained.

The **Mathcad** programs were written by Professor Jeffry Madura (Duquesne University) and they follow, to the extent that the syntax allow, the flow of the

Mathematica programs. All references to **Mathematica** programs in the text are also valid for the **Mathcad** programs.

I believe that one does not understand a theory if one is not able to turn it into a flow chart which can be translated into an algorithm. Because of this I have avoided providing programs where the student types some numbers in a box, clicks on a button and gets a table or a graph. Turning a theory into a working algorithm deepens the understanding of the theory and is an art that can be learned only through supervised practice. For this reason, the students are encouraged to write programs, for solving problems, rather than use "canned" software. Students can imitate the programs, which cover all the applications of the theory taught in the text, or use them for inspiration.

One advantage of using **Mathematica** or **Mathcad** is that it allows the use of more advanced theory and of real data, to connect classroom chemistry with laboratory practice. By presenting realistic calculations and comparing them with the experimental results, the Workbooks prepare students for their future professional lives.

The Workbooks also make it possible to work through realistic examples without the tedium of hand calculations. The students can focus on the physical chemistry phenomena, rather than the routine work of computation. By being able to process a large amount of data in milliseconds, they can study how phenomena change when the relevant parameters are modified. For example, one can calculate the equilibrium composition of complex reaction for many temperatures and pressures.

The use of **Mathematica** or **Mathcad** reduces the fear of mathematics: doing integrals, derivatives, power series expansions, taking limits, solving algebraic or differential equations become trivial.

Mathematica

Instructions for installing the **Mathematica** programs are provided in the ReadMe file in the CD-ROM. The Workbooks were created with **Mathematica** 5.1. The **Mathematica** tutorial is written to cover the needs of a physical chemistry student. The first two lectures teach the knowledge required for writing a primitive, working program. Subsequent lectures go into more details and show how to write safe programs that run efficiently. I provide these lectures on **Mathematica** because there is no "minimalist" **Mathematica** book, a shortcut to **Mathematica** directed to the needs of the physical chemist. This tutorial has been used several times as a textbook for a one-quarter course at the University of California, Santa Barbara for teaching **Mathematica** to chemistry and biochemistry students. Wolfram's book is available on the web, from inside any **Mathematica** program, and the reader can consult it when trouble arises. This exceptionally well-organized online help makes learning **Mathematica** a lot easier than any other computer language.

Professor Jeffry D. Madura's **Mathcad** Workbooks are a translation of the **Mathematica** Workbooks, to the extent that the syntax of the two languages allows this to be done. For all pedagogical purposes they are identical and the choice of one over the other should depend on availability, price, and instructor's preference.

Mathcad

A short introduction to **Mathcad** is provided as a PDF file on the CD-ROM to help the reader get started. This "tutorial" is also used to illustrate **Mathcad** features that are encountered in solving many of the problems. Each Workbook follows the **Mathematica** numbering scheme. That is, **Mathematica** notebook SM3.3 would be **Mathcad** Workbook 3.3. The Workbooks were written in **Mathcad** 12. To help those who do not have this version there are Adobe Acrobat PDF files of all the **Mathcad** Workbooks to view and use. All documents have been saved so that they can be read by older versions of **Mathcad**. The older versions have been placed in appropriately labeled folders and should be accessible in the same manner as described above.

Exercises and Problems

There is a trend, in recent textbooks, to include a large number of repetitive problems. I have resisted this trend. Once the calculation for ethane has been performed, nothing is gained by performing it for CO, CO_2, etc. I give detailed, solved exercises in the text out of the belief that a principle or methodology is understood only when one sees how it is implemented. These exercises are an integral part of the text and of the learning process. Unsolved exercises are placed throughout the chapters where they are most relevant to the book's content. This strategic placement reinforces what the reader is learning, as he or she learns it. Additional homework problems are included at the end of some chapters. Some exercises remain unsolved for a reason. The physical chemistry students are juniors or seniors, and many of them will go to work as scientists and engineers after they finish the course. It is a fair guess that the person asking them to calculate something does not know the answer. The students must learn how to test whether a result is reasonable. They must take responsibility for their work now, when the consequence for failure is a lower grade, not dismissal or tragedy. Every type of problem that a physical chemist needs to solve is illustrated, with a detailed example in the text and with even more details in the appropriate Workbook. This should offer sufficient guidance to students doing homework.

A few exercises have been included for which no data are given. Finding data is an essential part of the practice of chemistry and should be learned while in school. If a student is asked to calculate the change of enthalpy with temperature and is given the heat capacity and the enthalpy at 298 K, he or she is prompted to look for a formula that contains these two quantities. This crutch is absent in the working world. When asked to calculate the change of enthalpy with temperature, the student should know to look for a formula giving the derivative on enthalpy with temperature. Then, the student can figure out what data is needed for implementing the formula. Occasionally, I have asked the students to create a suitable homework for a given chapter. Presumably, a student that understood the material in a chapter should be able to figure out what he or she can do with it.

Below is a chapter-by-chapter listing of unsolved Exercises that can be assigned for homework:

Chapter One: 1.1, 1.2. *Chapter Two*: 2.1, 2.2, 2.3, 2.4, 2.5, 2.6, 2.7, 2.9, 2.10. *Chapter Three*: 3.1, 3.2, 3.3, 3.4, 3.5, 3.6, 3.7, 3.8, 3.9. *Chapter 4*: 4.1. *Chapter Five*: 5.1, 5.2, 5.3, 5.4. *Chapter Six*: 6.1, 6.2, 6.3, 6.4, 6.6, 6.7, 6.8, 6.9. *Chapter 7*: 7.1, 7.2, 7.3, 7.4. *Chapter Eight*: 8.1, 8.2, 8.3, 8.4, 8.5, 8.6, 8.7. *Chapter Nine*: 9.1, 9.2, 9.3, 9.4, 9.5, 9.6, 9.7, 9.8, 9.9, 9.10, 9.11, 9.12, 9.13, 9.14, 9.15, 9.16, 9.17, 9.18, 9.19, 9.20, 9.21, 9.22, 9.23, 9.24, 9.25, 9.26. *Chapter 10*: 10.1, 10.2, 10.3, 10.4, 10.5. *Chapter 11*: 11.1, 11.2, 11.3, 11.4, 11.5, 11.6, 11.7, 11.8, 11.9. *Chapter 12*: 12.1, 12.2, 12.3. *Chapter 13*: 13.1, 13.2, 13.3, 13.4, 13.5, 13.6, 13.7, 13.8, 13.9, 13.10, 13.11. *Chapter 14*: 14.1, 14.2, 14.3, 14.4. *Chapter 15*: 15.1, 15.2, 15.3

1

THE FUNDAMENTAL EQUATIONS OF STATISTICAL MECHANICS

Introduction

§1. *Why do you need Statistical Mechanics?* I will assume that you have already learned how to use thermodynamics to study large systems, such as a gas or a liquid. To develop that theory we introduced work, heat, energy, entropy, and formulated the First and Second Laws in terms of these quantities. Nowhere during this development did we need to know that the system is made of molecules. We invoked molecules occasionally, but their existence was never an essential part of the discussion. During the period in which the greatest developments in thermodynamics took place, some eminent physical chemists either denied the existence of molecules or dismissed the idea as an unnecessary hypothesis.

However, we now know that molecules exist and it is natural to ask how the quantities introduced in thermodynamics depend on their properties. For example, it would be nice to have equations that allow us to calculate the heat capacity or the equation of state, from the interactions among the molecules and their vibrational and rotational energies.

What can we gain from such a theory? Perhaps we can increase our understanding of thermodynamics in general, or of the thermodynamic properties of specific systems. For example, at high temperatures the heat capacity at constant volume, C_v, is equal to $3R/2$ for all noble gases (R is the gas constant). Why this number? Why are the noble gases different from other gases? Why does the heat capacity of a gas increase with the number of atoms in its molecules? Why are the C_v values of certain diatomics such as H_2 and HCl very similar and why do they differ from those of I_2 or Br_2? Thermodynamics cannot explain such similarities and differences. Statistical mechanics — which is a theory that connects the thermodynamic properties of a body to the properties of the molecules in it — can.

Phenomenological thermodynamics introduces entropy in a formal way and gives no physical interpretation of this central quantity. In some books, we are told that entropy is related to order: if a system is more ordered, its entropy is smaller. Where does this statement come from? What exactly is order? Why does entropy depend on it? Statistical mechanics manages to answer these questions.

When you studied chemical kinetics, you learned that a large number of experiments established that most rate constants, k, satisfy the Arrhenius equation $k = A\exp(-E_a/RT)$. Here A is the so-called pre-exponential, E_a is the activation energy, and R is the gas constant. Phenomenological chemical kinetics cannot tell us why this functional form is so common, or the meaning of the activation energy E_a or of the pre-exponential A. Nor can it tell us how k is related to the motion of the atoms in a molecule, or how the atoms in a molecule move during a chemical reaction. Statistical mechanics answers these questions.

We have a natural desire to understand how the world around us works. But, one could argue that a practical person can use thermodynamic data without understanding why the values of certain quantities have some striking regularities, or why entropy and order are related. After all, thermodynamics was a useful science before the invention of statistical mechanics.

It turns out that statistical mechanics has practical applications. A good example is provided by combustion reactions. These produce transient species (e.g. OH radicals) that are very reactive. We cannot isolate and purify them, to measure their thermodynamic properties, such as heat capacity, entropy, or energy of formation. Unless we know these properties, we cannot calculate how the temperature and the composition of a flame depend on the working conditions. This makes it harder to design boilers or engines that burn fuel efficiently and produce a minimum amount of pollutants. It so happens that it is rather easy to measure the emission spectrum of the OH radical in the flame and determine its vibrational and rotational energies. Using these quantities, statistical mechanics can calculate the thermodynamic properties of this transient species. Now the boiler makers can do their calculations. Much of the thermodynamic data from the JANAF database, which was developed to aid combustion computations, was calculated by using spectroscopic data and statistical mechanics.

In the past decade, computers have become so powerful that we can use statistical mechanics to calculate the thermodynamic properties of gases, liquids, solids, polymers, proteins, and enzymes. In some of these calculations, we are not limited by shortcomings of statistical mechanics or by lack of computer power, but by our incomplete knowledge of the energy of interaction between molecules. As our knowledge of these interactions improves, chemists, biochemists, and engineers will perform statistical mechanical calculations routinely to solve the problems that appear in their practice. As you read this, scientists at Merck, Pfizer, Schering-Plough, and other pharmaceutical companies are using computer programs based on statistical mechanics to help design better drugs.

§2. *Why is this Theory Statistical? A Discussion based on Classical Mechanics.* The premise of statistical mechanics is simple: the motion of the molecules in a material determines its macroscopic properties. Let us take this assertion seriously and think about its consequences. To make the argument simpler, assume that the motion of the molecules inside any material is described by classical mechanics and that the forces of interaction between the atoms or the molecules present in the system are known. Given these assumptions, I can use Newton's equation to calculate how each atom in the system moves. If the premise of statistical mechanics is correct, this should allow me to understand the macroscopic properties of the material.

This is a fine idea, but as I pursue it I will find that I don't know how to implement it. To solve Newton's equation, I must know the positions and the velocities of all the atoms in the system at one time. If I know that, then the position and the velocity of each atom at all other times can be calculated. Unfortunately, there is no experiment capable of giving me this information now or in the foreseeable future.

Even if the initial positions and velocities could be determined and Newton's equations solved, I do not know how to use this information to calculate the thermodynamic properties of the gas. I might assume that the energy of the system is the total mechanical energy of all the molecules in the system. But this cannot be quite right: I know that the energy depends on temperature, while the calculation just mentioned has no temperature in it.

Greater difficulties appear when I start thinking about temperature, heat, entropy, heat capacity, for example. Mechanics does not tell me how to use the positions and the velocities of every molecule in the gas to calculate its entropy or its temperature. To perform such calculations, I must add new concepts to mechanics, and this is what statistical mechanics does.

Finally, here is the most powerful blow to my initial idea: a theory that describes the motion of each atom in the system has no connection with thermodynamic experiments. In thermodynamics, I control the system very crudely, by fixing the temperature and the pressure. Let us call the state defined by giving the temperature T and the pressure p, the *macroscopic state* of the system. Thermodynamics tells me that this is enough to specify the state of a gas in equilibrium. In a mechanical description of the same system the state is specified by giving the momenta $\{\mathbf{p}_1, \ldots, \mathbf{p}_N\}$ and the positions $\{\mathbf{r}_1, \ldots, \mathbf{r}_N\}$ of the atoms. This is called the *microscopic state* $C = \{\mathbf{r}_1, \ldots, \mathbf{r}_N, \mathbf{p}_1, \ldots, \mathbf{p}_N\}$ of the system. To use mechanics, I must know the microscopic state of the system. However, our experience with thermodynamics says that it is enough to know T and p. Thus, there is a tremendous reduction of the amount of information as I go from mechanics to thermodynamics. Statistical mechanics must find a mathematical scheme for achieving this reduction; the theory has to show how to intelligently throw away most of the information contained in the microscopic state.

To do this, I start with the observation that there are a very large number of microscopic states compatible with a given macroscopic state. This means that there are many sets of different momenta and positions that will give the same pressure and temperature. Let us call these microscopic states, the microscopic states compatible with fixed T and p. When I fix the temperature and pressure, the system could be in any one of these states. Statistical mechanics recognizes that we have no way of determining in what microscopic state the system is. In the absence of this knowledge, the next best thing is to know the probability $P(C(T,p))$ that the system has a certain microscopic state $C(T,p) = \{\mathbf{r}_1, \ldots, \mathbf{r}_N, \mathbf{p}_1, \ldots, \mathbf{p}_N\}$. I use the notation $C(T,p)$ to indicate that this is a microscopic state for which the gas has the pressure p and temperature T. The probability $P(C(T,p))$ will depend on T and p.

Next I assume that the measured thermodynamic energy U is the average energy

$$U(T,p) = \sum_{C(T,p)} E(C(T,p))\, P(C(T,p)) \tag{1.1}$$

The quantity $E(C(T,p))$ is the mechanical energy of the system in the microscopic state $C(T,p)$. This is the kinetic energy of all atoms in the system, plus the energy of the interaction between them. The sum in Eq. 1.1 is over *all microscopic states compatible with the fixed values of T and p*. The whole expression is the average energy of all the molecules, at given values of temperature and pressure.

To define entropy, we need a new idea; mechanics alone is insufficient and statistical mechanics provides the necessary formula. Once we know the internal energy U and the entropy S, we can calculate all other thermodynamic quantities, by following the rules of thermodynamics.

Exercise 1.1

Go back and read your thermodynamics book and try to figure out how you can calculate all thermodynamic quantities if you know $U(T,p)$ and $S(T,p)$.

§3. *Why use a Statistical Method? A Quantum Mechanical Discussion.* The argument just given is based on the assumption that I can describe the motion of the molecule by using classical mechanics. This is often a good approximation, but the world of the molecules is ruled by quantum mechanics and a sound theory of statistical mechanics must be based on it. Statistical mechanics based on classical mechanics is sometimes an excellent approximation, but it is not an exact theory.

To describe the microscopic states of a quantum system, I must solve the time-independent Schrödinger equation. This gives me the energy eigenstates Ψ_i, and the corresponding energy eigenvalues, E_i, that the system can have. If the system were alone in the world, its state would be the lowest energy eigenstate Ψ_0. However, in thermodynamics the system is in contact with a thermostat. I know from quantum mechanics that the interaction of a system with something (in our case the thermostat) causes transitions from the initial energy eigenstate Ψ_0 to other eigenstates Ψ_i. For a very large system the energy difference between various energy eigenstates is very small. Because of this, the contact with the thermostat will cause transitions to a very large number of eigenstates.

Exercise 1.2

Get your quantum mechanics book and find the formula for the energy eigenstates of a particle in a box. Calculate the first 10 energy eigenstates for an argon atom enclosed in a cubic box of volume 1 cm^3. Find out how much energy it takes to excite the system from the ground state to the tenth excited state. Then calculate the change of the thermodynamic energy per atom when you increase the temperature by 10 K (from 298 K to 308 K). Use $C_p = 20.786$ J/mol K. Is this energy enough to excite the system from the ground state to the tenth excited state?

In a macroscopic experiment, I do not control the energy but instead the temperature and the pressure. There are many energy eigenstates that are compatible with a given pair T, p, and we do not know for sure what the microscopic state of the system is. Just as in the classical case, statistical mechanics provides the probability that for certain values for p and T, the system is in the microscopic state Ψ_i. With these probabilities, I can calculate the mean energy and the entropy of the system.

The Fundamental Equations

§4. *The Fundamental Equations are Guesses.* I will give here the basic equations of statistical mechanics for a system in *thermodynamic equilibrium.* They are not derived by logical arguments from other laws of nature that we take for granted; they were obtained by inspired, educated guesses. Certain arguments do lead us to believe that these guesses are reasonable, but they don't prove them. We believe that the equations are correct because mathematical deductions and computations based on them agree with a large number of results obtained by measurements. For example, the entropy and the specific heat of gaseous Xe calculated from these equations is in excellent agreement with the measured values, at all temperatures.

Since this is your first encounter with statistical mechanics, I will write down the basic equations without presenting the rationalization used to justify them. Such justifications are better understood after you have had some experience with the applications of the theory. One of the justifications is presented in the books by Terrell Hill or Donald McQuarrie (see the Further Reading at the end of the book).

§5. *The Fundamental Equations: the Probability that the System is in a Specific State.* Consider a system containing N molecules in thermodynamic equilibrium. They are enclosed in a box of volume V and are in contact with a thermostat that keeps the temperature equal to T. From quantum mechanics, I know that any system of N molecules in a box of volume V has an infinite number of energy eigenstates Ψ_α and the energy eigenvalues $E_\alpha(N, V), \alpha = 0, 1, 2, \ldots$. In this notation it is understood that $E_0 \leq E_1 \leq E_2 \leq \ldots$.

The energy $E_\alpha = E_\alpha(N, V)$ is an energy of the whole system (the whole gas or liquid). Calculating it is a problem in quantum mechanics. In principle, to do this I write the Hamiltonian H of the system, which is the sum of the kinetic energies of all atoms plus the interaction energy between them. Then, to obtain the energy eigenstates of the system, I solve the Schrödinger equation $H\Psi = E\,\Psi$. This solution gives me a list of energy eigenstates Ψ_α and the corresponding energies E_α. Here α is an integer, or a collection of integers, that labels the states.

Had the system been alone in the world, it would have slowly emitted radiation and gone into the lowest energy state. However, the interaction with the thermostat causes transitions from one energy eigenstate to another. Because of this, I no longer know what the state of the system is. However, statistical mechanics gives me the next best thing: the probability P_α that the system has the state Ψ_α, with energy E_α. This is given by

$$P_\alpha(N, T, V) = \frac{\exp[-E_\alpha(N, V)/k_B T]}{Q(N, T, V)} \tag{1.2}$$

with

$$Q(N, T, V) = \sum_{\alpha=0}^{\infty} \exp[-E_\alpha(N, V)/k_B T] \tag{1.3}$$

The sum Q is the *partition function* of the system. This is a central quantity in equilibrium statistical mechanics, and most of this book is dedicated to its calculation and use.

When applying these equations, it is necessary to distinguish between the states Ψ_α and the energies E_α. As you learned in quantum mechanics, it is possible that several states have *the same energy*. Such states are called *degenerate*. In most cases, there are physical reasons for such degeneracy. For example, the degenerate states of an electron in the hydrogen atom have the same energy but different angular momentum and spin.

It is important to remember that the sum in Eq. 1.3 is over all states. If there are three states having the same energy ε, the term $\exp(-\varepsilon/k_B T)$ appears three times in the sum defining Q. You should also realize that the probability that a molecule has a specific energy is different from the probability that the molecule is in a specific state having that energy. For example, if the states Ψ_7, Ψ_8, and Ψ_9 are degenerate and have the energy ε, then the probability that the system is in the state Ψ_7 is

$$\frac{\exp(-\varepsilon/k_B T)}{Q},$$

while the probability that the system has the energy ε is

$$\frac{3\exp(-\varepsilon/k_B T)}{Q}$$

The eigenvalues E_α depend on the volume because the particles are confined to a box of volume V. As you have learned in the lectures on quantum mechanics, the energy of a particle in a box depends on the volume of the box. The dependence of E_α on the number of particles N is easy to understand: very roughly speaking, the total energy is the sum of the energy of all molecules so more molecules means higher energy. The energies E_α do not depend on temperature.

The quantity k_B is Boltzmann's constant:

$$\begin{aligned} k_B &= 1.3806 \times 10^{-16} \text{ erg/K} = 0.695104 \text{ cm}^{-1}\text{/K} \\ &= 3.29977 \times 10^{-27} \text{ kcal/K} = 8.61771 \times 10^{-5} \text{ eV/K} \end{aligned} \tag{1.4}$$

§6. *The Fundamental Equations: Connection to Thermodynamics.* Eq. 1.2 is very useful, but it is not enough. To develop a molecular theory of thermodynamics, I must find a recipe for calculating thermodynamic properties. It can be shown that the Helmholtz free energy $A(N, T, V)$ is given by

$$A(N, T, V) = -k_B T \ln Q(N, T, V) \tag{1.5}$$

Eqs 1.2, 1.3, and 1.5 are the starting point of statistical mechanics. The rest of this book will teach you how to use them to study a variety of chemical systems and phenomena.

§7. *Formulae for other Thermodynamic Quantities.* You have learned when you studied thermodynamics that you can calculate any thermodynamic quantity from the Helmholtz free energy. I summarize below the equations used to perform such calculations. These relationships are used throughout this book and you might want to keep these pages handy. If you want to review how these equations are derived in thermodynamics, you may want to use my textbook. The lower case letters u, s, and h denote the energy, the entropy and the enthalpy of a mole of substance.

- The entropy S of the system is

$$S = -\left(\frac{\partial A}{\partial T}\right)_{N,V} \tag{1.6}$$

- The pressure p is

$$p = -\left(\frac{\partial A}{\partial V}\right)_{N,T} \tag{1.7}$$

- The internal energy U of N molecules can be calculated from either

$$U = A + TS \tag{1.8}$$

or

$$U = -T^2 \frac{\partial}{\partial T}\left(\frac{A}{T}\right)_{N,V} \tag{1.9}$$

- The enthalpy is

$$H = U + pV \tag{1.10}$$

- The heat capacity at constant volume is obtained from

$$C_v = \left(\frac{\partial u}{\partial T}\right)_{N,V} \tag{1.11}$$

or

$$C_v = T\left(\frac{\partial s}{\partial T}\right)_{N,V} \tag{1.12}$$

- The heat capacity at constant pressure is obtained from

$$c_p = \left(\frac{\partial h}{\partial T}\right)_{p,N} \tag{1.13}$$

or

$$c_p = T\left(\frac{\partial s}{\partial T}\right)_{p,N} \tag{1.14}$$

- The Gibbs free energy for N molecules is

$$G = A + pV = U - TS + pV = H - TS \tag{1.15}$$

The first equation is the definition; the second follows from the definition $A = U - TS$; the third is obtained by using the definition of enthalpy (which is $H = U + pV$).

- The chemical potential per molecule in a system with one component is

$$\mu = \left(\frac{\partial A}{\partial N}\right)_{T,V} \tag{1.16}$$

or

$$\mu = \frac{G}{N} \tag{1.17}$$

For a mixture, the chemical potential of component i is

$$\mu = \left(\frac{\partial G}{\partial n_i}\right)_{T,p,n_j} \tag{1.18}$$

where n_i is the number of moles of component i. The derivative is taken while keeping constant the temperature, the pressure, and all n_j other than n_i. The subscript T, p, n_j reminds us of this.

The quantities A, S, U, and H are extensive: they are proportional to the number of molecules in the system. If I take $N = N_A = 6.023 \times 10^{23}$ (this is Avogadro's number, which is equal to the number of molecules in 1 mole), then I obtain the values of these functions per mole. If I divide by N, I obtain the values per molecule.

§8. *Summary of the Recipe for Calculating Thermodynamic Functions.* To start, I must calculate the quantum energies E_α of the system. Then I can obtain Q from Eq. 1.3 and all thermodynamic quantities from Eqs 1.5–1.18. The only limitation in applying this theory is our inability to calculate the energies E_α for complicated systems. If such calculations could be performed reliably, the equations listed above would completely replace all thermodynamic measurements.

§9. *Classical Statistical Mechanics.* Many of the systems in which chemists are interested are described well by classical mechanics. In this case, Eqs 1.2 and 1.3 take a different form and the partition function is easier to calculate. The computational methods are so powerful that if we know the interactions between the atoms in the system, we can calculate thermodynamic properties for liquids, solids, polymers, proteins, etc. Unfortunately, the time alloted for statistical mechanics in the chemistry curriculum is very short and you will not learn here how these methods are used. As Baudelaire said, art is long and life is short.

2

THE PHYSICAL INTERPRETATION OF THE FUNDAMENTAL EQUATIONS OF STATISTICAL MECHANICS

Introduction

§1. This chapter and the next examines some consequences of the general formulae given in Chapter 1. Here I clarify what controls the magnitude of the probability P_α, that the state of the system is the energy eigenstate Ψ_α, and give a few examples of the use of this quantity. I will occasionally call P_α the probability that the system is in the state α, or the population of the state α.

I begin the chapter by reminding you of a few simple things about probability and by emphasizing that the concept of probability has predictive power only when we are interested in the results of a large number of experiments. This is often the case in statistical mechanics, which studies experiments on a large number of molecules. We either measure the average value of a quantity or we measure the fraction of molecules that display a certain behavior. In such situations, we can predict the outcome of the experiment by using probabilities. Not long ago people managed to perform measurements on a single molecule. These can be analyzed by using probabilities only if we repeat the measurement many times, under identical conditions.

This is followed by an analysis of the equation for P_α, which displays the fundamental role played by the energy k_BT: it is unlikely that we will find the system in a state whose energy is much larger than k_BT and it is quite probable to find the system in a state whose energy is less than or equal to k_BT. As far as the states of the system are concerned, the energy k_BT is the divide separating the haves from the have-nots.

In the last part of the chapter, I discuss experiments in which we measure the probability that a system is in a state Ψ_α having the energy E_α. The simplest system I could think of is a diatomic molecule embedded (dissolved) in a solid. Because of the confinement imposed by the solid, the molecule can only vibrate; its translational and rotational motions are hindered. By using Eq. 1.2, I calculate the probability that the molecule is in a vibrational state Ψ_α and show how these probabilities affect the emission and absorption spectra of this system. The understanding gained through this analysis led people to develop matrix isolation spectroscopy and night-vision glasses.

An Operational Definition of Probability

§2. Probability plays such a central role in statistical mechanics that it is worthwhile to remind you of a few simple things about it. They will help explain what kind of information is provided by Eq. 1.2.

Consider a set of N coin-flipping experiments, which produce heads n_h times and tails n_t times. The quantity

$$\nu_h = \frac{n_h}{n_h + n_t} = \frac{n_h}{N} \tag{2.1}$$

is called the *frequency* with which heads appear in the experiment. Tails appear with the frequency

$$\nu_t = \frac{n_t}{N} \tag{2.2}$$

Obviously,

$$\nu_h + \nu_t = 1$$

When I ask the students in my class how big n_h and n_t are, most answer: they are both equal to $N/2$; if no one is cheating and the coin is unbiased, half of the coin flips produce heads and half produce tails. If that is true, then $\nu_h = \nu_t = \frac{1}{2}$ (use Eqs 2.1 and 2.2).

This answer is not correct. To see this, let us perform a "coin-flip experiment" on the computer. Each computer language has a function that produces random numbers. For example, when the computer evaluates the function Random[Integer,{0,1}] provided by **Mathematica**, the result is either zero or 1, at random. Like a coin flip, this function is designed so that its two values appear with *equal probability*. (At least that was the intent of the designers. Generating truly random numbers is difficult, and the function Random in **Mathematica** is not among the best random number generators. However, I will assume that the function returns 0 or 1 randomly and equally often.)

If I use the function 100,000 times, I get a string (1,0,1,0,0,0,1,1 . . . , 1) of zeros and ones. I count the number of times 1 appears and denote it n_t; the number of times 0 appears is denoted n_h. With these values I can calculate the frequencies ν_h and ν_t from Eqs 2.1 and 2.2.

The results of several such computer experiments, performed in Workbook SM2.1, are shown in Table 2.1. Note that no experiment yielded an equal number of heads and tails. This means that $\nu_h \neq \nu_t \neq \frac{1}{2}$. Also note that as I increase the number of tosses from 10,000 to 1,000,000, ν_h and ν_t get closer and closer to $\frac{1}{2}$. I will therefore assume that when the number of flips goes to infinity, ν_h and ν_t tend to $\frac{1}{2}$. These limiting values of ν_h and ν_t, denoted by p_h and p_t, are called the *probabilities* that I obtain heads or tails in a particular coin flip.

§3. *We can use Probabilities only to Examine a Large Number of Random Events.* We have established that the probability to observe heads in the

Table 2.1 Results of the computer simulation of the coin-tossing experiment described in the text.

Number of "flips"	n_t	n_h	ν_t	ν_h
10,000	4,966	5,034	0.4966	0.5134
50,000	25,133	24,867	0.5027	0.4973
100,000	49,954	50,046	0.4995	0.5005
1,000,000	500,132	499,868	0.5001	0.4999

coin-flipping experiments is $\frac{1}{2}$. How do we use this information? Let us assume that someone plans to perform N coin flips and wants to know how many of them will produce heads. If we assume that N is a sufficiently large number then $n_h/N \equiv \nu_h \approx p_h = \frac{1}{2}$. Now I can answer the question: the number of times heads appears is $n_h \approx N/2$ (also $n_t \approx N/2$). But, we seem to have reached a contradiction here. In the computer experiment performed above, $n_h \neq n_t$ even when N was as high as 1,000,000. In fact n_h is never equal to n_t, except by a very infrequent accident. However, the relative error

$$\frac{n_h - N/2}{n_h}$$

contained in the statement $n_h \approx N/2$ becomes smaller and smaller as N increases. It is in this sense that the predictions made by probability theory become correct when N is large.

§4. In some cases I can use logic to calculate the probabilities. If the coin is fair, heads and tails are equally probable and $p_h = p_t$. By definition $\nu_h + \nu_t = 1$, so when the number of flips tends to infinity this relationship becomes $p_h + p_t = 1$. Together, these equations lead to $p_h = p_t = \frac{1}{2}$.

Exercise 2.1

(a) Use logic to determine the probability that you get 1, or 2, or 3, . . . , or 6 when you throw a fair die once. (b) Perform a computer experiment to simulate such throws. Determine the frequency and monitor how it converges toward the probability as the number of throws increases. *Hint*. Use the function Random[Integer,{1,6}].

Exercise 2.2

A roulette wheel has 36 slots, half red and half black. What is the probability that the ball stops at slot 20? What is the probability that it stops at a red slot? What kind of experiment would you perform to determine if the wheel is rigged?

Exercise 2.3

(a) You are playing a game in which you gain if a coin flip turns up heads and you lose if it turns up tails. The last five flips gave tails. Should you bet a large amount of money on the next flip? Explain your reasoning. (b) Simulate your strategy.

Generate a string of zeros and ones and find when there are three zeros or three ones in a row. Check if the flip after three zeros gives 1 more often than 0, or if it gives 0 more often than 1.

Exercise 2.4

You have heard that the probability that two unrelated terrorists bring a bomb on the same flight is extremely small. Is it a good strategy to carry a bomb with you every time you fly? Will this lower your chances of being blown up? This is an exercise in mathematics, which ignores the fact that you will not be blown up because you'll never make the flight; you'll end up in jail instead.

§5. *Fluctuations.* As you have seen above, even when the number of events (coin flips) is large, the number of heads n_h counted in experiment is never equal to the value $p_n N$ predicted by probability theory. The difference $n_h - p_h N$ is called a *fluctuation* in n_h. The fluctuation tells me how large an error I make when probability theory is used to calculate a quantity (such as n_h). The fluctuations themselves are random variables, meaning that if I do two coin-flip experiments, each performing N flips, I get two different values for the fluctuation of n_h. As explained in §3, the relative magnitude $(n_h - p_h N)/n_h$ of the fluctuation becomes smaller and smaller as N increases, but it could be sizeable when N is not very large.

Why bother with all this? Under certain conditions the fluctuations in certain quantities can be observed. Consider the case of light scattering by a liquid. According to thermodynamics the density of a liquid in equilibrium is the same at every place in the liquid. Now let us send a beam of light through such a uniform liquid. I can solve Maxwell's equations to find out how light will propagate. Without making any approximations, I find that it will move in a perfectly straight line. Experiments show that a large fraction of the light moves in a straight line, as the theory predicts, but a small amount goes in other directions. The theory is wrong!

Well, not quite. Remember that thermodynamics makes predictions regarding the *average values* of various properties. The correct statement in thermodynamics is that the average value of the density in the liquid must be the same at each point. Next, I will explain that in a real liquid the local density can differ from the average density and this difference causes light to deviate from the straight path. I define the *average density*, $\bar{\rho}$, to be the total number of molecules in the liquid divided by the total volume of the liquid. I will also define the *local density*, ρ, to be $n/\delta V$, where n is the number of molecules in a cube of volume δV (whose side is of length ~ 500 Å, about half the wavelength of light). The *average* number of molecules in the volume δV is $\bar{n} = \bar{\rho}\delta V$. If I could count the number of molecules in the volume δV, I would observe two things: first, the counted number of molecules n in δV differs from the average number of molecules $\bar{n}$, that is, the number of molecules in the small volume fluctuates; second, if I place the volume δV in two different places in the liquid, the numbers of molecules in the two volumes differ from each other. Therefore, the local density ρ depends on the location of the small volume in the liquid.

To summarize: observed on a length scale of the order of the wavelength of light, the density of a liquid is not uniform; the local density ρ at various places in the liquid differs slightly from the mean density $\bar{\rho}$, due to density fluctuations. Such small differences are legitimately ignored when we study thermodynamics. But we cannot ignore them when we study light propagation through the liquid. The direction of the movement of light depends on the refractive index of the medium, which is a function of density; the local index of refraction fluctuates as the density fluctuates. Light is sensitive to fluctuations in the index of refraction taking place on a length scale of about half the wavelength; the fluctuations cause a change in its direction of propagation. Because the fluctuations are small (the volume δV contains quite a large number of molecules), most light goes straight through the liquid; only a small fraction of the photons are diverted, as observed in experiments.

Statistical mechanics provides the probability that a fluctuation will occur, and we can calculate the probability that the local density has a certain value. Using the local density (instead of the average density) in Maxwell's equations, we can predict how much light is scattered in a certain direction, as a function of light frequency. Light-scattering experiments confirm that the predictions of this theory are correct. Moreover, the theory says that the probability that density fluctuations have a certain value depends on the compressibility of the liquid. This allows us to use light-scattering experiments to measure compressibility. The values obtained in this way are in excellent agreement with those measured directly by compressing liquids.

This example, and many others, show that fluctuations do exist and sometimes have measurable consequences.

§6. *What Does This Have to do with the Probability of Finding the System in a Certain State?* One way of interpreting the probability P_α that the system is in state Ψ_α is to use a Gibbs ensemble. This is an imaginary collection of $\mathcal{N}$ identical copies of the system that I want to study. 'Identical' here means that these copies have the same number of molecules N, held at the same temperature T and pressure p. In other words, they all have the same *thermodynamic (macroscopic) state*. As you learned in Chapter 1, fixing N, p, and T does not fix the *microscopic* state; many microscopic states are compatible with a given macroscopic state and a given copy could be in any one of these microscopic states. Now imagine that you could measure in which microscopic state each copy is. Let $\mathcal{N}_\alpha$ be the number of copies that happen to be in the state Ψ_α. If $\mathcal{N}$ is very large, the probability that a system is in the state Ψ_α is

$$P_\alpha = \frac{\mathcal{N}_\alpha}{\mathcal{N}} \tag{2.3}$$

The Gibbs ensemble is a mathematical fiction used to help us think more precisely about probability in statistical mechanics. It is the traditional way of presenting the subject. Another (equivalent) definition of P_α uses repeated measurements to determine the microscopic state of one system. After each measurement, we let the system go back to equilibrium and then repeat the measurement. I do this $\mathcal{N}$ times, and denote by $\mathcal{N}_\alpha$ the number of times that the measurement found the system in state Ψ_α. The probability P_α is then given by $\mathcal{N}_\alpha/\mathcal{N}$ if $\mathcal{N}$ is very large.

Truth in advertising requires me to tell you that such measurements cannot be performed in practice. A large system has a huge number of energy eigenstates whose eigenvalues are very close together. There is no experiment with sufficient energy resolution to determine the energy eigenstate of the system. The imaginary experiment just described serves only to give a concrete representation of P_α. It is possible, however, to measure P_α, as you will see shortly.

The Properties of P_α

§7. To derive some of the properties of P_α, I rewrite Eqs 1.2 and 1.3 in an equivalent, but more illuminating, form:

$$\begin{aligned} P_\alpha &= \frac{\exp[-E_\alpha/k_BT]}{Q} = \frac{\exp[-E_\alpha/k_BT]}{\sum_{\mu=0}^{\infty}\exp[-E_\mu/k_BT]} \\ &= \frac{\exp[-E_\alpha/k_BT]}{\exp[-E_0/k_BT]\left(1+\sum_{\mu=1}^{\infty}\exp[-(E_\mu-E_0)/k_BT]\right)} \\ &= \frac{\exp[-(E_\alpha-E_0)/k_BT]}{1+\sum_{\mu=1}^{\infty}\exp[-(E_\mu-E_0)/k_BT]} \end{aligned} \tag{2.4}$$

These equalities were obtained by factoring $\exp[-E_0/k_BT]$ in the partition function and then using $\exp[x]\exp[-x] = 1$, $1/\exp[-x] = \exp[x]$, and $\exp[-x]\exp[+y] = \exp[-(x-y)]$.

Eq. 2.4 tells me that the differences $E_\alpha - E_0$, not the individual energies, control the properties of P_α. This is as it should be: the zero of the energy scale can be chosen arbitrarily, but the values of P_α must not depend on that choice. You can easily see that if I add a constant C to each of the energies E_α (which means that I change the zero of the energy scale) the energy difference $(E_\alpha + C) - (E_0 + C)$ is not affected by this addition and neither is P_α.

Let us examine Eq. 2.4. Since the energy labeling is such that $E_{\alpha+1} \geq E_\alpha$, if follows that $E_\alpha - E_0 \geq 0$ for all values of α. This means that

$$0 \leq \exp\left[-(E_\alpha - E_0)/k_BT\right] \leq 1 \tag{2.5}$$

This implies that the denominator in Eq. 2.4 is always larger than 1; it is also independent of α.

Now let us look at the numerator of Eq. 2.4. Because of the way we labeled the energies we have

$$\exp\left[-\frac{E_0}{k_BT}\right] \geq \exp\left[-\frac{E_1}{k_BT}\right] \geq \exp\left[-\frac{E_2}{k_BT}\right] \ldots \tag{2.6}$$

Equality holds only if the two states being compared are degenerate. But this implies that

$$P_0 \geq P_1 \geq P_2 \ldots \tag{2.7}$$

with equality when the states are degenerate (e.g. $P_0 = P_1$ only if states 0 and 1 have the same energy).

We conclude that the larger the “quantum number” n of a state, the smaller its population and that the population of the ground state is always the highest. If we equate the energy of a molecule with its predisposition to activity, we must conclude that molecules prefer to be lazy; “*dolce far niente*” (delightful idleness) is their favorite state.

A look at Eq. 2.4 also leads us to conclude that a state α has a reasonable population if

$$\frac{E_\alpha - E_0}{k_BT} \leq 1 \tag{2.8}$$

Table 2.2 The values of $k_B T$ at several temperatures.

Temperature, T(K)	$k_B T$(eV)
100	0.00860
300	0.02585
500	0.04310
700	0.06030
900	0.77600

This displays the special role of the quantity $k_B T$ in statistical mechanics. It divides the states into two classes, those that are highly populated and those whose population is low.

Table 2.2 gives the values of $k_B T$ for various temperatures. These numbers tell me that at room temperature (roughly 300 K) the population of states whose energy exceeds that of the ground state by more than 0.025 eV is fairly low.

Exercise 2.5

Suppose that a molecule has two states, having the energies $E_\alpha = 0$ and $E_\beta = 0.3$ eV. Calculate P_α/P_β for $T = 100$ K, 500 K, 1000 K, and 2000 K. Calculate the energy E_β for which $P_\beta \leq 0.01$, at temperatures 300 K and 600 K.

Exercise 2.6

Calculate the energy $k_B T$, where T is temperature, in units of kcal/mole and J/mole, at the temperatures used in Table 2.2.

Exercise 2.7

Suppose that the energy E_α of the state α is 2 eV higher than the ground state energy. (a) Calculate the ratio P_α/P_0 when the molecule is kept at room temperature. (b) At what temperature would you have to keep the molecule if you want P_α/P_0 to be equal to 10^{-3}?

Exercise 2.8

Prove that $P_n/P_m = \exp[-(E_n - E_m)/k_B T]$.

A Few Examples of Populations and Their Use

§8. *Introduction.* In what follows I will use the theory developed above to calculate the populations of the states of a diatomic molecule embedded in a solid. Then I explain why such calculations are important for understanding the temperature dependence of emission and absorption spectra. At the end, I touch briefly on two applications in which we make use of the dependence of the population of various states on temperature.

§9. *A Diatomic Molecule Embedded in a Solid.* I examine a system consisting of many diatomic molecules of the same kind, embedded in a solid. Such a system can be prepared, for example, by bringing a very cold surface into contact with a gaseous mixture containing a lot of Ar and few diatomic molecules. The gases will freeze together on the surface, forming an Ar ice with the diatomic molecules embedded in it.

Because the molecules are trapped inside the solid, they cannot translate or rotate, but they have enough room to vibrate. This simplifies the analysis because we need only worry about the vibrational states of the diatomic molecules.

Quantum mechanics tells me that the vibrational energy is given by

$$E_{\alpha} = \hbar\omega\left(\alpha + \tfrac{1}{2}\right) - \hbar\omega x\left(\alpha + \tfrac{1}{2}\right)^{2}, \quad \alpha = 0, 1, 2, \ldots \tag{2.9}$$

This formula may differ from the one you learned in quantum mechanics: it includes a correction needed because the potential is not harmonic. Had we made the harmonic approximation, the energy would have been $\hbar\omega\left(\alpha + \frac{1}{2}\right)$, which ought to be familiar to you.

Because the molecule cannot rotate or translate, its vibrational energy is its total energy. Here I ignore the interaction of the diatomic molecule with the solid, to avoid making the situation too complicated. Taking this interaction into account will cause a slight shift of the vibrational frequency, which does not affect the conclusions reached here.

Consider that the diatomic molecules are the system, while the solid is a thermostat (one of its roles is to keep the temperature of the diatomic molecules constant). In statistical mechanics of systems at equilibrium, which we study here, the system has a given temperature, so there is always a thermostat somewhere, defining what the temperature is and keeping it constant.

The probability that a diatomic molecule inside the solid is in the vibrational state α is given by

$$P_\alpha = \frac{\exp\left[-E_\alpha/k_B T\right]}{Q} = \frac{1}{Q}\exp\left[-\frac{\hbar\omega\left(\alpha+\frac{1}{2}\right)-\hbar\omega x\left(\alpha+\frac{1}{2}\right)^2}{k_B T}\right] \tag{2.10}$$

with

$$Q = \sum_{\alpha=0}^{\infty}\exp\left[-\frac{E_\alpha}{k_B T}\right]. \tag{2.11}$$

I obtained this by using Eq. 2.9 for the energies that appear in the general equations, Eqs 1.2 and 1.3.

§10. *K_2 Molecules Inside a Solid.* To perform numerical calculations I use as an example the K_2 molecule. Spectroscopists have measured that for K_2, $\hbar\omega = 92.3\ \text{cm}^{-1}$ and $\hbar\omega x = 0.354\ \text{cm}^{-1}$. People working with infrared radiation prefer to use cm^{-1} as a unit of energy. I will convert this to eV by using the relationship $1\ \text{cm}^{-1} = 1.2398\times 10^{-4}$ eV. If E_α is in eV units, I must use the Boltzmann constant $k_B = 0.0000861771$ eV/K.

To calculate the probability that the K_2 molecule is in the state α, I use these data in Eqs 2.10 and 2.11:

$$P_\alpha = \frac{1}{Q}\exp\left[-\frac{1.2398\times 10^{-4}\left\{92.3\left(\alpha+\frac{1}{2}\right)-0.354\left(\alpha+\frac{1}{2}\right)^2\right\}}{8.61771\times 10^{-5}T}\right] \tag{2.12}$$

with

$$Q = \sum_{\alpha=0}^{\infty}\exp\left[-\frac{1.2398\times 10^{-4}\left\{92.3\left(\alpha+\frac{1}{2}\right)-0.354\left(\alpha+\frac{1}{2}\right)^2\right\}}{8.61771\times 10^{-5}T}\right] \tag{2.13}$$

These values of P_α for various states α and temperatures T are given in Table 2.3. They were calculated in Workbook SM2.3.

Table 2.3 shows that at very low temperatures, most of the molecules are in the ground state. As the temperature is increased, more and more high-energy states are populated. This can be easily understood through the relationship between E_α and $k_B T$, as discussed in §7.

Table 2.3 The probability $P_\alpha(T)$ of finding the K_2 molecule in the vibrational state α for several states and temperatures. The second column gives the energy of the state.

α	E_α (eV)	P_α (10)	P_α (40)	P_α (300)
0	5.77×10^{-3}	0.999998	9.64×10^{-1}	3.55×10^{-1}
1	1.72×10^{-2}	1.66×10^{-6}	3.46×10^{-2}	2.28×10^{-1}
2	2.86×10^{-2}	3.06×10^{-12}	1.28×10^{-3}	1.47×10^{-1}
3	3.99×10^{-2}	6.25×10^{-18}	4.82×10^{-5}	9.47×10^{-2}
4	5.11×10^{-2}	1.41×10^{-23}	4.87×10^{-6}	6.14×10^{-2}
5	6.22×10^{-2}	3.53×10^{-29}	7.43×10^{-8}	3.99×10^{-2}
6	7.33×10^{-2}	9.77×10^{-35}	3.03×10^{-9}	2.61×10^{-2}
7	8.42×10^{-2}	2.99×10^{-40}	1.27×10^{-10}	1.71×10^{-2}
8	9.50×10^{-2}	1.02×10^{-45}	5.44×10^{-12}	1.12×10^{-2}
9	1.06×10^{-1}	3.82×10^{-51}	2.40×10^{-13}	7.40×10^{-3}
10	1.16×10^{-1}	1.58×10^{-56}	1.08×10^{-14}	4.90×10^{-3}
11	1.27×10^{-1}	7.32×10^{-62}	5.01×10^{-16}	3.25×10^{-3}
12	1.38×10^{-1}	3.74×10^{-67}	2.38×10^{-17}	2.17×10^{-3}

Exercise 2.9

Make a family of plots of $P_\alpha(T)$ versus α, for several fixed values of T, and a family of plots of $P_\alpha(T)$ versus T, for several values of α.

§11. *Application: Emission Spectroscopy.* How can I verify that the populations P_α calculated in §10 correspond to reality? I know from quantum mechanics that a molecule in an excited state *emits a photon and goes into a state with lower energy* (if the transition is allowed). The frequency of the emitted photon is given by the famous formula $(E_f - E_i)/\hbar$, where E_i is the energy of the initial state (before the transition) and E_f is the energy of the final state (after the transition).

This formula can be applied to the vibrations of K_2. Quantum mechanics shows that the most probable transitions are those in which the vibrational quantum number changes by 1. The energies of the photons emitted in these transitions are

$$\hbar\omega_\alpha = E_\alpha - E_{\alpha-1} = \hbar\omega(1 - 2x\alpha) \tag{2.14}$$

Table 2.4 $E_\alpha - E_{\alpha-1}$ is the energy of the photon emitted from a K_2 molecule in state α. $P_\alpha(10)$ is the probability that the molecule is in the state α when the temperature is 10 K, and $P_\alpha(300)$ is the same quantity when the temperature is 300 K.

α	$E_\alpha - E_{\alpha-1}$ (cm^{-1})	$P_\alpha(10)$	$P_\alpha(40)$	$P_\alpha(300)$
1	92.5	1.66×10^{-6}	3.46×10^{-2}	2.27×10^{-1}
2	91.8	3.06×10^{-12}	1.28×10^{-3}	1.46×10^{-1}
3	91.1	6.25×10^{-18}	4.82×10^{-5}	9.46×10^{-2}
4	90.4	1.41×10^{-23}	1.87×10^{-6}	6.13×10^{-2}
5	89.7	3.53×10^{-29}	7.43×10^{-8}	3.99×10^{-2}
6	89.0	9.77×10^{-35}	3.03×10^{-9}	2.60×10^{-2}
7	88.2	2.99×10^{-40}	1.27×10^{-10}	1.71×10^{-2}
8	87.5	1.02×10^{-45}	5.44×10^{-12}	1.12×10^{-2}
9	86.8	3.82×10^{-51}	2.40×10^{-13}	7.39×10^{-3}
10	86.1	1.59×10^{-56}	1.08×10^{-14}	4.89×10^{-3}

The last equality is obtained by using Eq. 2.9 for E_α and $E_{\alpha-1}$. This differs from the formula given in most quantum mechanics books, because I have taken the effect of unharmonicity into account in Eq. 2.9.

The more common (but less accurate) harmonic approximation to the frequency of the emitted photon is obtained by taking $x = 0$ in Eq. 2.14. Notice that if you do that, the emission frequency no longer depends on α, which is contrary to observations.

The energy of a photon emitted by a molecule in state α is listed in Table 2.4, together with the probability $P_\alpha(T)$ that the state is populated at temperature T. The table tells me that a molecule in state $\alpha = 3$ emits a photon of energy 91.1 cm^{-1}. This emission is possible only if a molecule is in the state $\alpha = 3$, and the number of emitted photons is proportional to the number of molecules in that state. If the temperature is 10 K and the sample contains N molecules, only $P_\alpha N = 6.25 \times 10^{-18} N$ molecules will be in state $\alpha = 3$. In this case it would be difficult to detect the photons of frequency 91.1 cm^{-1} coming from the sample.

The situation is different when $T = 300$ K (we assume that the matrix does not melt at this temperature). In that case, there are many molecules in states $\alpha = 1$ to 10. The sample will emit photons of all the energies shown in the second column

of Table 2.4. If I plot the intensity of the emitted radiation versus its frequency, we will have peaks at the frequencies 92.5 cm^{-1}, 91.8 cm^{-1}, ..., 86.1 cm^{-1}. If $T = 10$ K, we will see at most a peak at 92.5 cm^{-1}.

The presence of the peaks is a qualitative confirmation of the theory and so is the fact that as the temperature is increased, new peaks are detected. I would like, however, to have a quantitative confirmation of the theory. This is also possible, because quantum mechanics shows that the intensity of the light emitted by a molecule in state α is proportional to the number of molecules $P_\alpha N$ in that state:

$$\mathcal{I}(\omega_\alpha) = C(\omega_\alpha) P_\alpha N$$

Here $C(\omega_\alpha)$ is a proportionality constant, ω_α is the frequency (energy) of the emitted photon, and $\mathcal{I}(\omega_\alpha)$ is the intensity of the emitted light. Quantum mechanics provides a formula for $C(\omega_\alpha)$.

The intensity $\mathcal{I}(\omega_\alpha)$ is roughly the height of the peak measured at the frequency ω_α. The ratio of two such peaks is

$$\frac{\mathcal{I}(\omega_\alpha)}{\mathcal{I}(\omega_\beta)} = \frac{C(\omega_\alpha) P_\alpha}{C(\omega_\beta) P_\beta} \tag{2.15}$$

The left-hand side of this equation can be determined by measuring $\mathcal{I}(\omega_\alpha)$ and $\mathcal{I}(\omega_\beta)$. The right-hand side can be calculated. If our theories (quantum mechanics, which gives $C(\omega_\alpha)$, and statistical mechanics, which gives P_α) are correct, then the measurements and the calculation ought to give the same result, and they do.

§12. *A Thermometer.* If Eq. 1.2 is used in Eq. 2.15, I find that

$$\frac{\mathcal{I}(\omega_\alpha)}{\mathcal{I}(\omega_\beta)} = \frac{C(\omega_\alpha)}{C(\omega_\beta)} \exp\left[-\frac{E_\alpha - E_\beta}{k_B T}\right] \tag{2.16}$$

Imagine now that I place in a system a molecule whose energy eigenvalues E_α ($\alpha = 0, 1, \ldots$) are known. I can then measure the emission intensities $\mathcal{I}(\omega_\alpha)$ and $\mathcal{I}(\omega_\beta)$ and calculate the temperature from Eq. 2.16. This suggests that emission spectra can be used to determine the temperature of a body.

You might think that this is the silliest and most complicated thermometer one can make. Consider though the task of measuring the temperature of a flame or a star. People who design rockets, car engines, or boilers in power plants do many experiments in order to optimize fuel consumption and diminish pollution. They need to know how the temperature of the flame changes when they change the burning conditions. A simple way is to measure the intensity of light emitted by several transitions, preferably from the same molecule. By using Eq. 2.16, one can determine the temperature of the flame *at the point where the light is emitted*. Note that $E_\alpha - E_\beta = \hbar\omega_{\alpha\beta}$ is the frequency of the emitted light, which is also measured.

§13. *Absorption Spectroscopy*. For a variety of reasons, we perform a measurement called absorption spectroscopy. We shine light on a sample and measure the intensity of the light that passes through. The absorption spectrum is a plot of the intensity of the transmitted light versus the frequency of the incident light. Such a plot has dips (minima) at certain frequencies. The number of dips and their depth depends on temperature.

Let us see whether we can understand why such an absorption spectrum depends on temperature. Consider a sample consisting of K_2 molecules embedded in a solid. Statistical mechanics tells us that at high temperature, many states of the molecule are populated. For example, the probability that a K_2 molecule is in the 12th excited vibrational state, at 300 K, is $P_{12}(300) = 0.00217$ (see Table 2.3). In a system that has N such molecules, $0.355N$ are in the ground state (energy E_0), $0.228N$ are in the first excited state, etc. (see Table 2.3).

What happens when we shine light through the gas? When the light frequency equals $(E_1 - E_0)/\hbar$, the molecules that are in the ground state will absorb it. Because of this, we record a dip in the intensity of light passing through the sample. The magnitude of the dip is proportional to the number of molecules P_0N in the ground state. At 300 K this number is $0.355N$, and at 10 K it is practically N. As we slowly change the light frequency ω, we find that absorption takes place also when $\omega = (E_2 - E_1)/\hbar$. A dip in light transmission is registered at this frequency. At 300 K its depth is proportional to $P_1N = 0.228N$, while at 10 K it is proportional to $1.66 \times 10^{-6}N$. At 300 K further dips are observed when ω is equal to $(E_3 - E_2)/\hbar$, $(E_4 - E_3)/\hbar$, etc. Their depths are proportional to $P_2N = 0.147N$, $P_3N = 0.095N$, etc. At 10 K, the depths of the same dips are proportional to $P_2N = 3.06 \times 10^{-12}N$, $P_3N = 6.25 \times 10^{-18}N$, etc. Obviously, there are fewer dips in the spectrum at low temperature.

We see that the depth of the dips in the absorption spectrum mirrors the magnitude of the probabilities P_α. By using statistical mechanics to calculate P_α, we can calculate the depth of the dips in the absorption spectrum. The results agree with the measurements, confirming the validity of the theory.

§14. *Matrix Isolation Spectroscopy*. The absorption spectrum of a gas at room temperature is very complicated. One reason for this is absorption by the molecules in the excited states. If we lower the temperature, the probability of finding a molecule in an excited state decreases and many of the dips in the absorption spectrum disappear. If we could manage to have practically all molecules in the ground state, then the spectrum would be easier to interpret since we can only have transitions $E_0 \rightarrow E_\alpha$; the transitions $E_1 \rightarrow E_\alpha$, $E_2 \rightarrow E_\alpha$, etc., are suppressed.

To bring all molecules into the ground state, we must have $E_1 - E_0 \gg k_B T$. This relationship can be satisfied for vibrational energies, but not for rotations. The difference $E_1 - E_0$ for rotational motion is so small that we cannot bring all molecules into the rotational ground state without a major effort. We can, however, eliminate rotations by freezing the molecule of interest in an ice of argon or another chemically inert molecule. This is done by blowing, on a very cold surface, a gaseous mixture of Ar atoms and the molecules whose spectrum we want to measure. The gases freeze together on the surface. Trapped in the ice, the molecule can no longer rotate and so photon absorption by excitation of rotations is no longer possible. The molecules have enough room to vibrate, but by cooling to 10 K we can bring them all to the ground state. At that point, only the vibrational excitation $E_0 \rightarrow E_1$ is possible and the spectrum is simple and easy to understand.

This method of obtaining absorption spectra is called matrix isolation spectroscopy. It has one more advantage besides the ones mentioned. Often we are interested in transient, reactive species that we cannot make in large amounts in a gas because they are chemically unstable. Often we can trap, accumulate, and store them in an Ar ice, and take their spectrum.

§15. *Night-Vision Glasses.* The theory says that regardless of the nature of the system, if we hold it at room temperature, there is some probability that the system is in an excited state. Quantum mechanics says that a system in an excited state will sooner or later emit a photon (this is an approximation: some molecules manage to get rid of their high energy by a "radiationless transition," which does not result in photon emission). Since your body is warm and is supposed to emit photons, why don't I see you in a dark room? Why is there dark, if all bodies around us emit photons? This question would be easy to answer if I were surrounded by Ar atoms (for example). The excitation energy of these atoms is so high that the probability of having an excited Ar atom around is minuscule. There would be no glow from the Ar surrounding me. However, your body is loaded with large molecules whose excitation energies are low. There is no doubt that, if the theory is right, many of these molecules are excited at room temperature and emit light. If this is true, why don't you glow in the dark?

The answer has to do with a peculiarity of our eyes and the magnitude of $k_B T$ at room temperature. Let me remind you of two things. The probability that the system is in an excited state with energy E_β is proportional to $\exp[-(E_\beta - E_0)/k_B T]$. At room temperature, $k_B T$ is approximately 0.025 eV (see Table 2.2). The probability that a state whose energy E_β is higher than the energy of the ground state by 0.025 eV (at room temperature) is proportional to $e^{-0.025/0.025} = 1/e$. This is a sizable probability and therefore the bodies around us will emit photons of this energy.

If that is so, why don't I see this radiation? It so happens that the human eye uses certain molecules to detect light and these are not activated by photons of such low frequency. The glow of your body would be visible if it were capable of emitting light of higher frequency. But this is not likely: the probability that the energy of a molecule is 1 eV above the ground state (and can emit a photon with that energy), at room temperature, is less than $e^{-1/0.025} = e^{-40} = 4.248 \times 10^{-18}$. This is an extremely small number. So, I conclude that you glow in the dark, and emit a lot of photons of low frequency that I cannot see and an extremely small number of photons that could be detected by my eye.

But let us assume now that we are in the dark and I have a device that detects low-frequency photons. If that device clicks when it detects a low-frequency photon, then I would hear a lot of noise when it is oriented to receive radiation from your direction. I can hear your glow!

Can I use this observation to make a rudimentary device to detect you in the dark? A bit of thought will quickly discover a difficulty. This would be a fine instrument if you were the only glowing thing in the room. But the walls, the floor, the ceiling and everything else glows just like you. My detector will make noise no matter how I orient it. This is almost right, but not quite. If it is a cold night outside, your body is likely to be warmer than the walls. According to the theory, it will have more excited molecules and will emit more light. My detector will make more noise when I direct it towards you. I can use it to find where you are.

This idea, combined with some very sophisticated electronics that transforms the optical signal into an image (not into noise), led to the construction of night-vision glasses. These glasses see the temperature of various surfaces: the higher the temperature of a surface, the brighter its image. The resolution is not good enough to allow me to admire your beauty, but it is sufficient for a soldier who wants to shoot you in the dark. It has been claimed that the British won the Falklands War in 1982 because they had better night-vision equipment. It is doubtful that this wonderful natural law is in place to make it easier to shoot people, but such are the laws of unintended consequences.

3

INTERPRETATION OF THE THERMODYNAMIC QUANTITIES

§1. *Introduction.* In the previous chapter we studied one of the fundamental quantities of statistical mechanics, the probability that the system is in a given quantum state. Here we learn what statistical mechanics has to say about thermodynamic quantities such as energy, pressure, work, and heat exchange between the system and the medium and entropy. We are particularly interested in understanding the connection between macroscopic quantities and the properties of the molecules in the system.

Here is a summary of what you will learn. We will start with the mechanical quantities, energy and pressure. We will find that the equations of statistical mechanics imply that the thermodynamic energy is the average of the quantum energies of the system and the pressure is the average of the force exerted on the unit area of the wall.

You might think that this is a trivial result, but it isn't. The average value of a quantity x is an expression of the form $\sum_{i=0}^{\infty} x_i p_i$, where the symbols x_i represent all possible values of x, and p_i is the probability that x takes the value x_i. It is easy to say that the macroscopic energy studied in thermodynamics is the average of the microscopic energies, but it is not easy to say what probability one should use in the equation. The same comment is valid for pressure. Moreover, the energy and the pressure are derived in thermodynamics in a very abstract way (as I remind you below). It is very reassuring that the abstract formulae of statistical mechanics and the abstract formulae of thermodynamics lead, when combined, to equations that agree with our common sense.

While one can have some intuitive opinions about mechanical quantities, no one can possibly have an intuitive understanding of entropy, after having studied phenomenological thermodynamics. You have heard, perhaps, that entropy is linked to disorder: the higher the disorder, the higher the entropy. But what is disorder? This is a rather vague concept. Statistical mechanics makes this concept quantitative, by showing that entropy is a measure of the information we have about the system. The more information we have, the more order there is. The clothes in my closet are ordered if I know where every garment is, that is, if I have complete information about the system. There is less information, and less order, if I know that the pants I want to wear are somewhere in the left half of the closet. It turns out that one of the formulae for entropy, provided by statistical thermodynamics, is a quantitative measure of the quality of the information we have about the system.

§2. *The Starting Point.* The connection between the microscopic properties of a system and its thermodynamic quantities is made by the following two equations (these were given in Chapter 1)

$$Q = \sum_{\alpha=0}^{\infty} \exp\left[-\frac{E_\alpha}{k_B T}\right] \tag{3.1}$$

$$A = -k_B T \ln Q \tag{3.2}$$

Q is called the partition function of the system. The sum defining it is over all possible *quantum states* of the system, which are labeled by $\alpha = 0, 1, \ldots, \infty$. E_α are the energies of these states. Note that we sum over all states, not over all energies; if there are ten states having the same energy E, the exponential $\exp[-E/k_B T]$ appears ten times in the sum.

A is the Helmholtz free energy that you studied in thermodynamics. As you learned there, if you know how A depends on the temperature T, volume V, and the number of particles N, then you can calculate all the other thermodynamic functions. The equations needed for doing this have been collected in Chapter 1 (see Eqs 1.6–1.18).

Here is our task: start with Eqs 3.1 and 3.2 and then use Eqs 1.6–1.18 to derive expressions for all thermodynamic quantities. Then generate, from these equations, some understanding of these abstract thermodynamic quantities in terms of the microscopic properties of the system.

Energy

§3. Thermodynamics defines the energy U of a system very abstractly. The First Law postulates that the function U, defined by

$$dU = \delta W + \delta B, \tag{3.3}$$

is a function of state called energy. Here δW is the work performed on or by the system in an infinitesimal transformation (i.e. we change the state (that is, pressure and temperature) by a small amount) and δB is the heat exchanged between the system and the medium in the transformation. The symbol δW is used, rather than dW, to emphasize that δW is not an exact differential (as U is).

I apologize for using B as a symbol for heat. The normal usage is Q but I am using this for the partition function. It is hard to find a letter that is not already used for something else; humankind is facing a shortage of symbols, besides those of oil and drinking water.

By the end of the nineteenth century some very distinguished physicists believed that molecules did not exist or, if they existed, physical chemistry does not need them: thermodynamics was enough. Other people believed that molecules do exist, and that the thermodynamic energy U is some sort of average of some sort of molecular energy. But average of what quantity, with what kind of probability? If Eqs 3.1 and 3.2 are correct, then they should be able to answer this question. You will see shortly that they do.

To show this I start with Eq. 1.9, which is

$$U = -T^2 \frac{\partial}{\partial T}\left(\frac{A}{T}\right)_{N,V} \tag{3.4}$$

Introducing $A = -k_B T \ln Q$ in Eq. 3.4 and performing the derivative gives

$$U = \frac{k_B T^2}{Q}\left(\frac{\partial Q}{\partial T}\right)_{N,V} \tag{3.5}$$

(I have used $\partial \ln x/\partial x = 1/x$ and the chain rule $\partial \ln f(x)/\partial x = \partial \ln f/\partial f \cdot \partial f/\partial x = (1/f(x))(\partial f(x)/\partial x)$.)

The derivative of Q with respect to T is

$$\frac{\partial Q}{\partial T} = \sum_{\alpha=0}^{\infty} \frac{\partial}{\partial T} \exp\left[-\frac{E_\alpha}{k_B T}\right] = \sum_{\alpha=0}^{\infty} \exp\left[-\frac{E_\alpha}{k_B T}\right] \frac{\partial}{\partial T}\left(-\frac{E_\alpha}{k_B T}\right)$$

$$= \sum_{\alpha=0}^{\infty} \frac{E_\alpha}{k_B T^2} \exp\left[-\frac{E_\alpha}{k_B T}\right]$$

In these calculations, I used the chain rule $\frac{\partial}{\partial x}\exp[-f(x)] = \exp[-f(x)]\frac{\partial}{\partial x}(-f(x))$ and $\frac{\partial}{\partial y}(1/y) = -1/y^2$. Inserting this expression for $\partial Q/\partial T$ into Eq. 3.5 gives

$$U = \sum_{\alpha=0}^{\infty} E_\alpha \frac{\exp\left[-E_\alpha/k_B T\right]}{Q} \tag{3.6}$$

But $\exp[-E_\alpha/k_B T]/Q$ is the probability P_α that the system is in the quantum state labeled α (see Eq. 1.2), and this turns Eq. 3.6 into

$$U = \sum_{\alpha=0}^{\infty} E_\alpha P_\alpha \tag{3.7}$$

We found that the thermodynamic energy U is the average value of the energy eigenstates of the system. It is very rewarding that the abstract, and very general, equations 3.2–3.7 lead, after a few mathematical manipulations, to an equation in agreement with our common sense.

Exercise 3.1

Use the data given in Chapter 2 to calculate the energy U of 1 mole of K_2 molecules trapped in a solid matrix. Plot the dependence of U on temperature, in the range $T \in [10\text{ K}, 500\text{ K}]$. Use units of cal/mol for energy and the data given in §10 of Chapter 2.

Pressure

§4. In thermodynamics we defined pressure when we introduced work: it is the force exerted on the wall of the system, per unit area. After we introduced the First and the Second Law we showed that (see Eq. 1.7)

$$p = -\left(\frac{\partial A}{\partial V}\right)_{N,T} \tag{3.8}$$

Since force = pressure × area, we can take Eq. 3.8 to be a very general definition of force. For example, if I grab the ends of a folded protein chain and increase the distance between them, I will be exerting a force. I can calculate this force by taking the derivative of the free energy A of the protein with respect to the distance and multiplying by the distance (distance replaces area here). It would be extremely difficult to define this temperature-dependent force by using mechanics.

Since a thermodynamic system is made up of molecules, it is very tempting to believe that the pressure is the average force exerted by the molecules on the unit area of wall. But, what is this force and what probability do we use to calculate the average? This is the same question we asked when we studied energy and we will try to answer it by using the same procedure.

I start with Eq. 3.8, use Eq. 3.2 for A and Eq. 3.1 for Q, and then I use calculus to evaluate the resulting expressions. I obtain:

$$\begin{aligned} p &= -\frac{\partial}{\partial V}\left(-k_B T \ln Q\right) = \frac{k_B T}{Q}\left(\frac{\partial Q}{\partial V}\right)_{N,T} \\ &= \frac{k_B T}{Q}\sum_{\alpha=0}^{\infty}\left(-\frac{1}{k_B T}\right)\left(\frac{\partial E_\alpha}{\partial V}\right)_N \exp\left[-\frac{E_\alpha}{k_B T}\right] \\ &= \sum_{\alpha=0}^{\infty} -\left(\frac{\partial E_\alpha}{\partial V}\right)_N P_\alpha \end{aligned} \tag{3.9}$$

I used $\partial \ln f(x)/\partial x = (\partial f/\partial x)(1/f(x))$, $\partial \exp f(x)/\partial x = \exp(f(x))(\partial f/\partial x)$, and Eq. 1.2 for P_α.

The subscript N, in the derivative $(\partial E_\alpha/\partial V)_N$, tells me that the number of molecules in the system is held constant when I take the derivative with respect to V. In other words I calculate how E_α changes if I change the volume infinitesimally and I do not allow molecules to enter the container or escape from it.

This formula tells me how to calculate the pressure, if I know the microscopic energies E_α of the system, their degeneracy, and their volume dependence. To understand the meaning of this formula, I need to know the physical meaning of $(\partial E_\alpha/\partial V)_N$. This is what I do next.

From mechanics we know that the component of the force vector in the x-direction is the derivative of the energy with respect to x, taken with a minus sign. Therefore, when a system is in the state α and has the energy E_α, it exerts the force

$$F_\alpha(x) = -\left(\frac{\partial E_\alpha}{\partial x}\right)_N \tag{3.10}$$

on a wall perpendicular to the x-direction. If I take the x-direction to be perpendicular to one of the walls of a cubic container and divide Eq. 3.10 by the area a, I obtain the pressure p_α, exerted on the wall when the system is in the state α:

$$p_\alpha \equiv \frac{F_\alpha}{a} = -\left(\frac{\partial E_\alpha}{\partial ax}\right)_N = -\left(\frac{\partial E_\alpha}{\partial V}\right)_N \tag{3.11}$$

To obtain the second equality, I use $dV = a\,dx$, where dV is the change in volume when the wall is displaced a distance dx.

If I use this definition of p_α in Eq. 3.9, I obtain

$$p = \sum_\alpha P_\alpha p_\alpha = \sum_\alpha P_\alpha \left(-\frac{\partial E_\alpha}{\partial V}\right)_N \tag{3.12}$$

Again, the formal derivations based on statistical mechanics lead to results in agreement with common sense: the thermodynamic pressure p is the average of the mechanical pressure p_α exerted on the wall when the system is in the microscopic state α. It is gratifying that we now have a precise rule telling us what microscopic quantity we average, with what probabilities.

§5. *The Dependence of E_α on N and V.* Before proceeding, let us ask ourselves why E_α would depend on N and V. The dependence on N is obvious: each molecule has a certain internal (vibrational, rotational) and translational energy; the more molecules in a box, the bigger the energy of the system.

Exercise 3.2

The statement in §5 concerning molecules in a box may seem to imply that the mechanical energy of the system is the sum of the energies of each molecule.

Is this statement true? Is there any system in which this statement is correct? Give a specific example.

The dependence on the volume is more subtle. You might remember from your studies of quantum mechanics that the energy of a particle in a cubic box is given by

$$E_{n,j,k} = \frac{\hbar^2 \pi^2}{2mL^2}\left[n^2 + j^2 + k^2\right] \tag{3.13}$$

You can find this equation in Chapter 8 of my book on quantum mechanics. There I give the formula for a parallelepiped with sides L_x, L_y and L_z; for a cube with side L, we have $L_x = L_y = L_z = L$.

Eq. 3.13 gives the energy we have denoted here E_α. It so happens that each α is a set of three numbers n, j, and k, which can take any of the values $n = 1, 2, \ldots$; $j = 1, 2, \ldots$; $k = 1, 2, \ldots$.

Since L is the side of the cube, $L^2 = V^{\frac{2}{3}}$ and the energy depends on the volume of the box. Since all thermodynamic systems are enclosed in some container, we conclude that their energy eigenvalues must depend on the volume (not necessarily with the dependence given above).

A further dependence on the volume appears because the molecules interact with each other. This interaction depends on the distance between them; if we change the volume, we change this distance and the interaction energy changes with it.

Exercise 3.3

Use Eq. 3.13 to calculate the force exerted on the walls by an electron enclosed in a cubic box, when the electron is in a specific quantum state. For all the calculations requested below, except for (d), use the states $n = 1, j = 1, k = 1$ and $n = 1, j = 1$, $k = 20$, and the cube sides $L = 1$ cm and $L = 10$ Å.

(a) Calculate the force exerted on the walls by one molecule.

(b) Divide the force by the area to calculate the pressure exerted by a molecule on the wall.

(c) Calculate the work needed to change the volume by five percent.

(d) Calculate the pressure by using Eq. 3.9. Perform this calculation only for the box having $L = 10$ Å. Try to understand why I didn't ask you to calculate the pressure for the box having $L = 1$ cm.

Exercise 3.4

This is a bit more difficult. The energy of a particle inside a parallelepiped, whose sides have the lengths L_x (along the x-direction) L_y, and L_z, is given by

$$E_{n,j,k} = \frac{\hbar^2 \pi^2}{2m} \left[\left(\frac{n}{L_x} \right)^2 + \left(\frac{j}{L_y} \right)^2 + \left(\frac{k}{L_z} \right)^2 \right], \tag{3.14}$$

with $n = 1, 2, \ldots; j = 1, 2, \ldots; k = 1, 2, \ldots$. Perform the same calculations as in the previous exercise but for a parallelepiped with the sides $L_x = 10$, $L_y = 30$ and $L_z = 20$. In addition, calculate the force needed to keep the wall perpendicular to the x-direction from moving (from changing L_x), when the particle is in the state $n = 1, j = 1, k = 1$. Do the same calculation for the y- and z-directions. Are these forces equal? If they are not, how do you explain the fact that the thermodynamic pressure, at equilibrium, is the same on all walls?

Work and Heat

§6. In thermodynamics, a change dU of the energy is intimately connected to the work δW and heat δB given to or taken from the system. The First Law says that

$$dU = \delta B + \delta W = \delta B - p\, dV \tag{3.15}$$

(I used $\delta W = -p\, dV$.) The internal energy is given by $U = \sum_{\alpha=0}^{\infty} E_\alpha P_\alpha$ (Eq. 3.7) and the change of U is

$$dU = \sum_{\alpha=0}^{\infty} E_\alpha\, dP_\alpha + \sum_{\alpha=0}^{\infty} dE_\alpha\, P_\alpha \tag{3.16}$$

I will use Eqs 3.15 and 3.16 to establish a connection among work, heat, and the molecular properties of the system.

Eq. 3.16 tells me that there are two ways of changing the thermodynamic energy U: I change the energy eigenstates E_α or the probability P_α (that the system is in a state α). Since the energy $E_\alpha(N,V)$ is a function of the number of particles N (which is held constant in this analysis) and the volume V,

$$dE_\alpha = \left(\frac{\partial E_\alpha}{\partial V}\right)_N dV$$

Introducing this in Eq. 3.16 leads to

$$dU = \sum_{\alpha=0}^{\infty} P_\alpha \left(\frac{\partial E_\alpha}{\partial V}\right)_N dV + \sum_{\alpha=0}^{\infty} E_\alpha \, dP_\alpha \tag{3.17}$$

According to Eq. 3.9, the first term in Eq. 3.17 is

$$\sum_{\alpha=0}^{\infty} P_\alpha \left(\frac{\partial E_\alpha}{\partial V}\right)_N dV = -p dV = \delta W \tag{3.18}$$

The last step follows from the definition $\delta W = -p\, dV$ of the work (given by thermodynamics).

Introducing Eq. 3.18 in Eq. 3.17 gives $dU = \delta W + \sum_\alpha E_\alpha \, dP_\alpha$. When I compare this to Eq. 3.15, I see that the heat exchanged between the system and the external world is

$$\delta B = \sum_{\alpha=0}^{\infty} E_\alpha \, dP_\alpha \tag{3.19}$$

I summarize now the results of this analysis. There are two ways to change the energy of a system in thermodynamics: by performing work or by exchanging heat. In statistical mechanics, work corresponds to a change of the microscopic energy eigenstates of the system

$$\delta W = \sum_{\alpha=0}^{\infty} P_\alpha \, dE_\alpha \tag{3.20}$$

Heat corresponds to a change in the probabilities that the system is in various states (see Eq. 3.19).

Work, which is a mechanical quantity, has a simple mechanical interpretation: the infinitesimal work exerted on the system is the average infinitesimal change in the microscopic energies of the system. As always, a "mechanical" thermodynamic quantity in thermodynamics is the average of a microscopic mechanical quantity. Heat is a "thermal quantity" and it is different: the heat exchanged in an infinitesimal transformation is not the average of some microscopic quantity. It is the change of energy caused by a change in the probability that the system has a certain energy. Since in a system in equilibrium $\delta B = TdS$ where S is the entropy, we will obtain a better understanding of heat later, when we understand entropy.

Exercise 3.5

From thermodynamics, you know that the heat δb needed to change the temperature of 1 mole of substance from T to $T + dT$ and its volume from V to $V + dV$ is $\delta b = C_v dT + \ell_v dv$, where v is the molar volume. Use Eq. 3.19 to derive equations for C_v and ℓ_v. Show that $C_v > 0$. Is there any physical reason why C_v must be positive?

Entropy

§7. Of all thermodynamic quantities, entropy is the most mysterious. The Second Law defines entropy as a *function of state* whose change during an *equilibrium transformation* is $dS = \delta B/T$. Most introductory textbooks go on to explain that somehow entropy is connected to disorder: the higher the disorder, the higher the entropy. For example, a liquid is more disordered than a solid, and therefore the entropy of liquid ammonia should be higher than that of solid ammonia. And indeed, it is. Thermodynamics has no way of connecting entropy to disorder, and statements like these often do not define precisely what disorder means, or how it is measured, or calculated. We feel that indeed a liquid is more disordered than a solid, and a gas is more disordered than a liquid. But if pressed, we probably won't be able to say precisely what that means.

§8. *A New Equation for Entropy.* Statistical mechanics provides a way of linking entropy to order, which is discussed here. As preparation for this discussion, it is necessary to derive a new equation for entropy. To do this I start from a known thermodynamic equation (the definition of Helmholtz free energy A): $A = U - TS$, from which I have

$$S = \frac{U - A}{T} \tag{3.21}$$

I know that $U = \sum_\alpha E_\alpha P_\alpha$ (Eq. 3.7) and $A = -k_B T \ln Q$ (Eq. 3.2). Inserting these two equations in Eq. 3.21 gives

$$S = \sum_\alpha \frac{E_\alpha P_\alpha}{T} + k_B \ln Q \tag{3.22}$$

Next, I take the logarithm of the equation $P_\alpha = \exp[-E_\alpha/k_B T]/Q$ (Eq. 1.2) and obtain

$$E_\alpha = -k_B T(\ln Q + \ln P_\alpha) \tag{3.23}$$

(I used $\ln(1/a) = -\ln a$ and $\ln \exp a = a$.) Finally, I insert this expression for E_α in Eq. 3.22 to obtain

$$S = \left[-k_B \sum_\alpha P_\alpha \ln P_\alpha - k \ln Q\right] + k_B \ln Q \tag{3.24}$$

which leads to

$$S = -k_B \sum_{\alpha=0}^{\infty} P_\alpha \ln P_\alpha \tag{3.25}$$

This is the equation I need in what follows. It is very remarkable that in this expression, the entropy depends only on the population of the states; no other property of the system is relevant.

§9. *What Do I Mean by Order?* I will now explain what this expression has to do with order. Earlier, I made the statement that if disorder is increased in a system, its entropy goes up. To make this meaningful, a quantitative measure of disorder is needed. I will show that the entropy given by Eq. 3.25 is such a measure.

The concepts of disorder, order, and information are related: if my library is perfectly well organized, then every book is cataloged and is in place, and I can say that it is perfectly ordered. This also means that I can find where every book is. I have perfect information about the system (the library). So order is in some sense connected to information. If the order is not perfect (not all the books are in their appointed places), then I cannot be absolutely sure that I can find a given book. I may be lucky and the particular book is where it is supposed to be and I can find it right away. But perhaps the book I want is one of the misplaced books, and in this case I cannot find it easily. Lack of order leads to poor information. Our first step is to accept that disorder is linked to information: more disorder means less information.

The next step is to establish that information is linked to my ability to guess the outcome of my actions. Let us assume that I have 100 books in my library and five of them are out of place. I don't know which ones are misplaced. If someone asks me for a book, I cannot be sure that I can find it on my first try. I can only say that the probability of finding it on my first try is 0.95. If more books are misplaced — and I don't know which — then the disorder is greater, the information is diminished, and the probability of finding a book is smaller. There is a connection among disorder, information, and probabilities.

Now that I have linked disorder to information and to probabilities, I will show that the entropy defined by Eq. 3.25 is a measure of the amount of information I have about the system. To see why I say this, I examine a few simple examples.

If I toss a coin, I have a fairly high chance of guessing the outcome. I can only get heads or tails, and the probability of each is 1/2. Compare this to throwing a die. There are now six equally probable outcomes and the probability for each of them is 1/6. There is more a priori information about the outcome of flipping a coin than about the outcome of tossing a die. This means that there is a better chance of guessing the outcome of a coin flip than that of a die toss.

Now let us calculate the entropy of these two systems, according to Eq. 3.25. For the coin flip there are two events: heads or tails. This means that the system has two states: heads, with probability 1/2, and tails, with probability 1/2. The two probabilities are equal because the coin is fair and they are 1/2 because they must add up to 1. The entropy of the coin flip is, according to Eq. 3.25,

$$S_{\text{Coin}} = -k_B[0.5\ln(0.5) + 0.5\ln(0.5)] = +0.693147k_B$$

I calculate next the entropy for tossing a die. The result can be any one of the numbers 1, 2, 3, 4, 5 or 6. Therefore the system has six states, and each has probability 1/6 (they are equal and they must add up to 1). The entropy, according to Eq. 3.25, is equal to

$$S_{\text{Die}} = -k_B \left[\overbrace{\tfrac{1}{6}\ln\tfrac{1}{6} + \cdots + \tfrac{1}{6}\ln\tfrac{1}{6}}^{\text{six times}} \right]$$

$$= -k_B \times 6 \left[\frac{1}{6} \ln \frac{1}{6} \right] = +1.79176 k_B$$

The entropy of the die toss is higher than that of the coin flip. The example confirms that the less information I have about the system, the higher is its entropy.

I can consider another example: I flip a coin that is rigged so that heads come up less often than tails. Suppose that the probability of getting tails is 0.75 and that of getting heads is 0.25. I have now more information about the system than when I used a fair coin: my ability to guess the outcome is increased (this is why a crook wants to use a rigged coin). The entropy for this system is

$$S_{\text{RiggedCoin}} = -k_B[0.75 \ln 0.75 + 0.25 \ln 0.25] = +0.562335 k_B$$

The entropy of the rigged coin flip is lower than that for the fair coin, in agreement with the statement that the more information I have (which is the same as more order in the system), the smaller the entropy.

You can amuse yourself with more examples. I mention in passing that this definition, inspired from statistical mechanics, is used to give a quantitative measure for information in fields that have nothing to do with physics.

§10. *The Connection to Statistical Mechanics.* What does this have to do with thermodynamics? Statistical mechanics is a sort of coin flip or die toss. I am not asking for the probability that we get heads or tails, but for the probability P_α that the system is in a given state α. If $P_0 = 0.9$ and $P_1 = 0.1$ and all other states have zero probability, then I have quite a bit of information about the states of the system. The entropy of the system is then low. If, however, $P_0 = 0.6$, $P_1 = 0.2$, $P_2 = 0.1$, $P_3 = 0.06$, $P_4 = 0.04$, and the probabilities of the other states are zero, then I have less information about the state of the system. The system is more "disordered" and its entropy is higher.

§11. *The Dependence of Entropy on Temperature.* It is common to claim, in introductory textbooks, that the entropy of a system increases as we increase the temperature, because a hot system is more disordered than a cold one. If you look at an entropy table for any substance, you will find that the entropy does increase with temperature. However, the vague statement given above does not explain what is meant when we say that the hot system is more disordered.

Now that we understand "disorder" and "information" more precisely, I will show that, indeed, the information I have about the system deteriorates when I increase the temperature. I will examine an example, but you will see from the calculation that this statement is very general. Let us consider the K_2 molecule examined earlier and calculate its entropy at 40 K and at 100 K. The molecule has the energy eigenstates $\hbar\omega(\alpha + \frac{1}{2})$ (I use here the harmonic approximation, for simplicity) and the probability that the molecule is in state α is

$$P_\alpha = \frac{\exp[\hbar\omega(\alpha + \frac{1}{2})/k_B T]}{Q}$$

with

$$Q = \sum_{\alpha=0}^{\infty} \exp\left[\frac{\hbar\omega(\alpha + \frac{1}{2})}{k_B T}\right]$$

Here are the values needed for evaluating these expressions: $\hbar\omega = 93.2\ \text{cm}^{-1}$, $k_B = 8.61771 \times 10^{-5}$ eV/K, and $1\ \text{cm}^{-1} = 1.2398 \times 10^{-4}$ eV.

Workbook

At 10 K, the probabilities are (see Workbook SM3.1) $P_0 = 0.999998$, $P_1 = 1.66364 \times 10^{-6}$, and $P_2, P_3, \ldots$ are smaller (remember that P_α decreases with α). When $T = 10$ K, I have a lot of information about this system; I can guess with a very high chance of success the state of the system. With odds like these in the stock market, I would easily become rich. The entropy at this temperature, calculated with Eq. 3.25 and these values of the probabilities, is rather small (see Table 3.1).

At 100 K the probabilities are $P_0 = 0.73$, $P_1 = 0.19$, $P_2 = 0.051$, $P_3 = 0.014$, $P_4 = 0.0038$, $P_5 = 0.0010$, $P_6 = 0.00029$, $P_7 = 0.8 \times 10^{-4}$, $P_8 = 0.2 \times 10^{-4}$, $P_9 = 6.67 \times 10^{-6}$, etc. It is still most likely that the system is in the ground state, but now the odds of guessing the state of the system are smaller. The information deteriorates when temperature is increased. The entropy, calculated with Eq. 3.25, is higher (see Table 3.1).

Exercise 3.6

Show that entropy grows with temperature regardless of the nature of the system.

Table 3.1 The temperature dependence of the entropy of one K_2 molecule embedded in a solid, as a function of temperature.

Temperature, T (K)	S/k_B
100	0.778
200	1.420
300	1.810
400	2.100
500	2.330
600	2.530
700	2.710
800	2.860
900	3.000
1000	3.130

As T increases we are less and less able to guess the state of the system; we have less information and the entropy increases. Since less information means less order, one can say that at a higher temperature the system is more disordered and has higher entropy.

Exercise 3.7

Compare two systems. One has the energies $E_\alpha = 0.1$ eV, 0.2 eV, 0.3 eV, and 0.4 eV; the other has the energies 0.5 eV, 0.7 eV, and 0.8 eV.

(a) Without performing any calculation, decide which system has higher entropy. Explain your decision.

(b) Use $S = -k_B \sum_\alpha P_\alpha \ln P_\alpha$ to calculate the entropy for the two systems at several temperatures between $T = 10$ K and $T = 500$ K. Which system's entropy varies more strongly with temperature?

(c) Perform the same analysis and calculations for two systems, one with the energy eigenstates 0.1 and 0.7 eV and the other with 0.1, 0.2, 0.3, and 0.4 eV.

(d) One system has the energies 0.1 and 0.2 eV and the other 0.1, 0.2, and 0.3 eV. Explain, without performing calculations, which one has higher entropy. Calculate the entropy to verify your prediction.

Exercise 3.8

Calculate the thermodynamic quantites per mole a, u, h, s, μ, C_p, and C_v for a K_2 molecule in a solid matrix. Use the data given in the text, and units of eV. Perform numerical calculations for $T = 10$ K, 100 K, and 1000 K. Explain the result you obtain for the pressure.

The Third Law of Thermodynamics

§12. When you studied thermodynamics you learned that Nernst had postulated, after analyzing a vast set of data, that the entropy of all systems is zero when the temperature reaches 0 K. Later, people found some exceptions: for example, the entropy per molecule of a CO crystal is $k_B \ln 2$, not zero. This result was mysterious at the time when the Third Law was new and was being tested. Let us see if statistical mechanics can derive Nernst's law and explain the deviations from it.

Let us accept, for the moment, the statement that at 0 K the system is in the ground state. This means that $P_0 = 1$ and $P_1 = P_2 = \ldots = 0$. The entropy of this system is

$$S = -k_B(1\, ln(1) + 0\, ln(0) + 0\, ln(0) + \ldots).$$

Since $\ln(0) = -\infty$ the terms $0 \ln(0) = 0(-\infty)$ are a bit troublesome. However, if I use L'Hôpital's rule (check your calculus book), I obtain $\lim_{P\to 0} P \ln P = 0$. **Mathematica** can be used to calculate this limit (see Workbook SM3.2) and gives the same result.

We can complete the proof that the entropy is zero when $T = 0$ K if we can show that at zero temperature $p_0 = 1$ and all other p_α are zero.

I start with

$$P_\alpha = \frac{\exp[-E_\alpha/k_B T]}{Q}$$

and

$$Q = \sum_\alpha \exp\left[-E_\alpha/k_B T\right]$$

I can write P_0 as

$$P_0 = \frac{\exp[-E_0/kT]}{\exp[-E_0/kT]\{1 + \exp[-(E_1 - E_0)/kT] + \exp[-(E_2 - E_0)/kT] + \cdots\}}$$

$$= \frac{1}{1 + \exp[-(E_1 - E_0)/kT] + \exp[-(E_2 - E_0)/kT] + \cdots}$$

By definition, $E_1 > E_0$, $E_2 > E_0$, etc., and therefore $E_1 - E_0 > 0$, $E_2 - E_0 > 0$, etc. Because of this, as $T \to 0$, I have $\exp[-(E_\alpha - E_0)/kT] \to 0$ for all $\alpha > 0$, and hence $P_0 \to 1$. Because $P_0 + P_1 + P_2 + \cdots = 1$, the fact that $P_0 \to 1$ when $T \to 0$ means that $P_\alpha \to 0$ for $\alpha > 0$. This is what we set out to prove.

The entropy at 0 K is zero because this is the most highly organized state of the system: we are certain to find the system in the ground state every time we measure the energy. You can understand now why physical chemists say that it is not possible to have a temperature lower than 0 K.

Now that I know why $S = 0$ when $T = 0$, I must also understand why experiments show that for a CO crystal, the entropy per molecule is $k_B \ln 2$. In the crystal, the CO molecule can be in one of two orientations, and each orientation is equally probable. If I label the two orientations by the indices 1 and 2, then $P_1 = 1/2$ and $P_2 = 1/2$. The entropy corresponding to this uncertainty in orientation is given by

$$\frac{S}{k_B} = -(P_1 \ln P_1 + P_2 \ln P_2) = -\left(\frac{1}{2}\ln\frac{1}{2} + \frac{1}{2}\ln\frac{1}{2}\right) = +\ln 2$$

Deviations from the Third Law can also come from the nuclear spin. Quantum mechanics tells me that if the nuclear spin of the molecule is I, then the ground state degeneracy is $2I + 1$. This means that I have $2I + 1$ states whose energy equals the ground state energy E_0 and which differ through the orientation of the nuclear spin. In the absence of a magnetic field, these states are equally probable. I denote the probability of one such state by $\theta = 1$ or $\theta = 1/(2I + 1)$. The entropy corresponding to these states is

$$S/k_B = -\sum_{i=1}^{2I+1} \theta \ln \theta = -\sum_{i=1}^{2I+1} \frac{1}{2I+1} \ln\left(\frac{1}{2I+1}\right) = +\ln(2I+1)$$

I have used here $\ln(1/x) = -\ln x$. This contribution to entropy is temperature-independent so it does not go to zero when $T \to 0$. Therefore the entropy at 0 K is $S = k_B \ln(2I + 1)$, causing a violation of the Third Law.

In general, if the degeneracy of the ground state is g_0, then $S/k_B = \ln g_0$ at 0 K.

As indicated earlier, the Third Law was derived by extrapolating a large body of data to lower and lower temperatures. In most cases S approached 0, but sometimes it did not. The deviations from the law were a puzzle and could not be understood within thermodynamics. The law follows trivially from statistical mechanics and so does the understanding of the deviations. Had statistical mechanics been invented before the Third Law was formulated, it would have saved a lot of work.

Exercise 3.9

Determine the values of a, u, h, s, μ, C_p, and C_v for a K_2 molecule in a solid matrix when $T \to 0$.

4

THE PARTITION FUNCTION OF A SYSTEM OF INDEPENDENT PARTICLES

§1. *Introduction.* The equations given in Chapter 3 allow me to calculate the thermodynamic properties of any system if I know how to determine its energy eigenvalues E_i. This requires solving the Schrödinger equation for the whole system, a task that we are not very good at. But life is kind to new theories and offers one system for which such calculations are easy. We know how to calculate the allowed energies E_i of a gas made of simple particles (atoms or small molecules) that do not interact with each other. Larger molecules give us trouble, because they have a complicated internal motion.

Once I know the energy eigenvalues E_i of the gas, I can calculate its partition function and thermodynamic properties, by following the recipe given in Chapter 3. I show here, by implementing that recipe, that the partition function of a system of independent particles is a product of partition functions of the N individual molecules present in the system. Next I postulate that, when all N particles in the system are identical, the product thus obtained needs to be divided by $N!$. In Chapter 5, I implement the equations obtained here and calculate the partition function of a gas of independent atoms. In Chapter 6, I derive formulae for the thermodynamic functions of such a gas. These are compared with the results of measurements in Chapters 7 and 8.

§2. *The Partition Function of a System of Independent Particles.* I study a box of volume V, held at temperature T, and containing N identical independent particles. By "independent" I mean that each particle behaves as if the others do not exist; each acts as if it is alone in the box. Since particles become aware of each other through interactions, they are independent when they don't interact.

Of course, all particles interact with each other and saying that they don't needs some explanation. It so happens that the interaction between neutral molecules falls off rapidly with the distance between the particles; it is extremely small at a distance larger than the "range of interaction", which is about 10 Å. As we decrease the density of the system, the average distance between the particles can become larger than the range of interaction. This means that at any given time the interaction energy between most particles is zero. Furthermore, each particle has a considerable kinetic energy, more so at higher temperatures. This means that if the temperature is high and the density is low (i.e. the pressure is small), the kinetic energy of a particle is so much higher than the energy of its interaction with the other particles that we can neglect the latter. This is what the phrase "the particles do not interact" means.

A particle that does not interact with its neighbors behaves as if it is alone in the box. I will label the possible states of a single particle in the box by i and denote their energies by ε_i. Moreover, ε_0 is the ground state and $\varepsilon_0 \leq \varepsilon_1 \leq \varepsilon_2 \leq \ldots$.

If the particles do not interact, the energy E_α *of the gas* is *the sum of the energies of the individual particles*:

$$E_\alpha = \varepsilon_{i_1} + \varepsilon_{i_2} + \cdots + \varepsilon_{i_N} \tag{4.1}$$

Here ε_{i_1} is the energy of particle 1 in state i_1, and i_1 could be equal to 0, 1, 2, ...; ε_{i_2} is the energy of particle 2 in state i_2, and i_2 could be equal to 0, 1, 2, ...; etc.

To specify the state of N particles, I specify the state i_1 of particle 1, the state i_2 of particle 2, etc. For example, the lowest energy state of the gas is described by the "multiple index" $\alpha = \{i_1 = 0, i_2 = 0, \ldots, i_N = 0\}$, when all particles are in the ground state (that of lowest energy); the energy of the gas in this state is $\varepsilon_0 + \varepsilon_0 + \cdots = N\varepsilon_0$. One excited state of the whole gas is described by $\alpha = \{i_1 = 1, i_2 = 0, \ldots, i_N = 0\}$ in which the first molecule is excited to the state $i_1 = 1$ and the rest are in the ground state; the energy of the gas in this state is $\varepsilon_1 + (N-1)\varepsilon_0$.

To calculate the partition function $Q = \sum_\alpha \exp[-E_\alpha/k_BT]$, I must sum over all states α; but this means that I must sum over all possible states i_1 of particle 1, all states i_2 of particle 2, etc.

Using Eq. 4.1 and this rule for summation leads to

$$Q = \sum_\alpha \exp\left[-\frac{E_\alpha}{k_BT}\right] = \sum_{i_1}\sum_{i_2}\cdots\sum_{i_N} \exp\left[\frac{-(\varepsilon_{i_1} + \varepsilon_{i_2} + \cdots + \varepsilon_{i_N})}{k_BT}\right] \quad (4.2)$$

Here k_B and T are, as usual, the Boltzmann constant and the temperature (in kelvin).

Because $\exp[x + y] = \exp[x]\exp[y]$, I can write Eq. 4.2 as

$$Q = \left(\sum_{i_1} \exp\left[\frac{-\varepsilon_{i_1}}{k_BT}\right]\right)\left(\sum_{i_2} \exp\left[\frac{-\varepsilon_{i_2}}{k_BT}\right]\right)\cdots\left(\sum_{i_N} \exp\left[\frac{-\varepsilon_{i_N}}{k_BT}\right]\right) \quad (4.3)$$

Since the particles are identical I have

$$\sum_{i_1} \exp\left[\frac{-\varepsilon_{i_1}}{k_BT}\right] = \sum_{i_2} \exp\left[\frac{-\varepsilon_{i_2}}{k_BT}\right] = \cdots = \sum_{i_N} \exp\left[\frac{-\varepsilon_{i_N}}{k_BT}\right] \equiv q \quad (4.4)$$

These equalities hold because the index i_1 takes the same values as i_2, etc., and the energy ε_{i_1} takes the same values as ε_{i_2}, etc.

Eq. 4.4 implies that I can write Eq. 4.3 as

$$Q = q^N \quad (4.5)$$

where

$$q = \sum_{i=0}^{\infty} \exp\left[-\frac{\varepsilon_i}{k_BT}\right] \quad (4.6)$$

Here ε_i are the energy eigenstates of one particle occupying the box alone. The sum in Eq. 4.6 is over all the states (not all energies) of such a particle. This means that if three states have the same energy ε, $\exp[-\varepsilon/k_BT]$ appears three times in the sum. In what follows I will call q the *one-particle partition function.*

§3. *If the Total Energy is a Sum, the Partition Function is a Product.* Look back at the mathematical manipulations that took us from Eq. 4.1 to Eq. 4.5. They allow us to make a general statement: if the energy is a sum of terms, such as $E_\alpha = \epsilon_i + \eta_j + \lambda_k$, the total partition function

$$Q = \sum_\alpha \exp\left[\frac{-E_\alpha}{k_B T}\right]$$

is a product of partition functions

$$Q = \sum_i \exp\left[\frac{-\epsilon_i}{k_B T}\right] \sum_j \exp\left[\frac{-\eta_j}{k_B T}\right] \sum_k \exp\left[\frac{-\lambda_k}{k_B T}\right]$$

Each term in the product is the partition function corresponding the energies ϵ_i, η_j, and λ_k.

This property is general and I will use it several times in this book.

§4. *A Correction for Identical Particles.* In spite of the impeccable logic used in deriving it, Eq. 4.5 is not correct. The partition function for N particles is not q^N (see Eq. 4.5) but

$$Q = \frac{q^N}{N!} \tag{4.7}$$

In this expression, $N!$ is the factorial of N:

$$N! = 1 \times 2 \times 3 \times \cdots \times (N-1) \times N$$

Historically, this correction was introduced in an ad hoc manner: without $N!$, the equation would disagree with experiment, and with it, the equation gives good results. It was a patch, an afterthought, a mysterious fix that the experiments foisted upon us. Later, as people understood quantum mechanics better, they managed to prove that $N!$ should be present in the equation if the N particles are strictly identical.

The reasons for the presence of $N!$ are subtle and we do not have time to do them justice. I will tell you what the conclusions are. You have learned, in your quantum mechanics course, that the electron wave function must be *antisymmetric.*

This is true for all fermions, which are particles whose spin is not an integer. Particles with integer spin are called bosons and their wave function must be *symmetric*. Because of these differences in their quantum mechanical properties, the statistical mechanics of fermions is different from that of bosons, and they both differ from the version presented here.

A careful analysis finds that if the particles are heavy, their density is low, and the temperature is sufficiently high, then the statistical mechanics of bosons becomes identical to that of the fermions and it is given by Eq. 4.7. The only remnant of the symmetry requirements is the division by $N!$.

What are the practical consequences of all this? It so happens that Eq. 4.7 is applicable to all gases, liquids, and molecular solids, except for He or hydrogen at very low temperature. You should be aware of these limitations, but you should not let them bother you. The theory works for most systems of interest to chemists. It is, however, worthless if you want to study electrons in a metal (which are fermions) or the photons in equilibrium with a hot body (which are bosons). In such cases the difference between fermions and bosons is substantial and these systems need to be treated with a more elaborate theory, which is not discussed here.

To remind us of the approximation used in deriving Eq. 4.7, we say that a system for which Eq. 4.7 is accurate "satisfies Boltzmann statistics".

§5. *Stirling's Approximation.* Having $N!$ in Eq. 4.7 may cause you to panic. If you are dealing with one mole of substance, $N = 6.022 \times 10^{23}$ (Avogadro's number). Calculating $1 \times 2 \times 3 \times \cdots \times (6.022 \times 10^{23})$ is not exactly a picnic; you would be very old by the time you performed all those multiplications. Fortunately, Stirling discovered a wonderful formula: if N is large enough (and 10^{23} surely is), then

$$N! \simeq \left(\frac{N}{e}\right)^N \tag{4.8}$$

Here $e \approx 2.71828$ is the base of the natural logarithm.

Exercise 4.1

Verify numerically that this equation is a better and better approximation as N increases.

Using Stirling's formula, I can write Eq. 4.7 as

$$Q = \left(\frac{qe}{N}\right)^N \tag{4.9}$$

Make sure that you remember that e and N inside the brackets, originate from division by $N!$ (the "patch"). It is interesting to keep track of the effect this correction has on the results of our calculations.

§6. *Is the Free Energy of a System of N Noninteracting Molecules Equal to N Times the Free Energy of One Molecule?* If we ignore the factorial and use Eq. 4.5 to calculate the free energy of the system, we obtain (start with Eq. 1.5)

$$A = -k_B T \ln(Q) = -k_B T N \ln(q) \tag{4.10}$$

Since q is the partition function of one molecule, the quantity $a \equiv -k_B T \ln(q)$ is the free energy of one molecule. According to Eq. 4.10, the free energy of a gas of N molecules that do not interact with each other is N times the free energy of one molecule. This seems eminently reasonable: the molecules do not interact and each molecule behaves as if it is alone in the box; the total free energy should be the sum of the free energies of a molecule.

The trouble with this result is that we have obtained it by using the wrong formula, Eq. 4.5. What happens when I use the correct formula, Eq. 4.9? The free energy is

$$A = -k_B T \ln(Q) = -k_B T N \ln(q) - k_B T N \left[\ln(N) - 1\right] \tag{4.11}$$

Let us try to understand this equation. If I divide the free energy A of the gas by the number N of molecules in it, I obtain the free energy *per molecule*:

$$\frac{A}{N} = -k_B T \ln(q) - k_B T \left[\ln(N) - 1\right] \tag{4.12}$$

Furthermore, the free energy in a system containing *one molecule* in the box is (make $N = 1$ in Eq. 4.7 and calculate $A = -k_B T \ln Q$)

$$-k_B T \ln(q) \tag{4.13}$$

When I compare Eq. 4.12 to Eq. 4.13 I find a very strange result: the free energy per molecule, in a gas of N noninteracting particles, differs from the free energy of one molecule. Why is this strange? It violates my premise that each molecule behaves as if it is alone in the box! The difference $-k_bT\,[ln(N)-1]$ originates from $N!$, which we introduced earlier.

A second strange fact about the free energy per particle is that *it depends on the number of molecules N in the gas*. This is possible only if the molecules interact! I am forced to conclude that the term $N!$ introduces some kind of interaction between the particles, even though *the energy of the interaction between particles is zero*, by definition.

This "interaction" is forced upon us by the requirement that the wave function of a system of N identical particles must be either symmetric or antisymmetric (this is why we introduced $N!$ in the theory). There is much to say about this, but this goes beyond the introductory level of this text.

This symmetry/antisymmetry requirement is one of the most remarkable discoveries of quantum mechanics. It has no classical analog and it has extraordinary consequences. If a miracle should change natural law and suppress this requirement, our world would change beyond recognition.

5

THE PARTITION FUNCTION OF AN IDEAL GAS OF ATOMS

§1. *Introduction.* The results derived in the previous chapter hold for a gas of particles that do not interact. Here I specialize these equations for a gas of non-interacting *atoms.* As a pedagogue I am interested in this system because of its simplicity: I can apply statistical mechanics without making any approximations. If the theory disagrees with the experimental results, we have no place to hide. We cannot claim that perhaps the difference between the theoretical prediction and the measurement is caused by the approximations made in deriving the equations and that the theory will perhaps be perfect once these approximations are removed.

The theory of a gas of non-interacting atoms is also of interest to practicing chemists. Atomic species are present in high-temperature flames and to design better combustion conditions (less pollution, appropriate temperature, fuel efficiency) we need to know their thermodynamic properties. Since atoms tend to be very reactive (noble gases are an exception), it is difficult to measure their properties. Fortunately, we can calculate them by using the equations derived in this chapter.

The energy of an atom is the energy of its nucleus, plus that of its electrons, plus that of its translational motion in the box in which it is enclosed. Because the energy of the atoms is a sum of terms, its partition function is *a product* (see Chapter 4, §3) of a nuclear, an electronic, and a translational partition function. Here I show how these three partition functions are calculated. Once the partition function of the gas is known, I can calculate all its thermodynamic properties, by using the prescription given in Chapter 3. I do this in Chapter 6. Comparison with experiments is performed in Chapters 7 and 8.

§2. *The Partition Function of an Atom is the Product of a Translational, a Nuclear, and an Electronic Partition Function.* The partition function of a gas of N non-interacting particles is given by (see Eq. 4.9)

$$Q = \left(\frac{qe}{N}\right)^N \tag{5.1}$$

where q is the partition function of one particle:

$$q = \sum_\alpha \exp\left[-\frac{\varepsilon_\alpha}{k_B T}\right] \tag{5.2}$$

In this equation ε_α is the energy of one particle, alone in the box.

An atom is not a point particle: it is made of electrons and has a nucleus. Quantum mechanics tells me that its energy is a sum of three terms

$$\varepsilon_i = \varepsilon^{\mathrm{n}}_\alpha + \varepsilon^{\mathrm{e}}_\beta + \varepsilon^{\mathrm{t}}_\gamma \tag{5.3}$$

Here $\varepsilon^{\mathrm{n}}_\alpha$, $\alpha = 0, 1, 2, \ldots$, are the energies of the nucleus; $\varepsilon^{\mathrm{e}}_\beta$, $\beta = 0, 1, 2, \ldots$, are the energies of the electrons; and $\varepsilon^{\mathrm{t}}_\gamma$, $\gamma = 0, 1, 2, \ldots$, are the energies of the translational motion of one atom alone in the box.

The partition function of an atom is therefore

$$\begin{aligned} q &= \sum_i \exp\left[-\frac{\varepsilon_i}{k_B T}\right] \\ &= \sum_\alpha \sum_\beta \sum_\gamma \exp\left[-\frac{\left(\varepsilon^{\mathrm{n}}_\alpha + \varepsilon^{\mathrm{e}}_\beta + \varepsilon^{\mathrm{t}}_\gamma\right)}{k_B T}\right] \\ &= \sum_\alpha \exp\left[-\frac{\varepsilon^{\mathrm{n}}_\alpha}{k_B T}\right] \sum_\beta \exp\left[-\frac{\varepsilon^{\mathrm{e}}_\beta}{k_B T}\right] \sum_\gamma \exp\left[-\frac{\varepsilon^{\mathrm{t}}_\gamma}{k_B T}\right] \end{aligned} \tag{5.4}$$

I call

$$q_{\mathrm{n}} = \sum_{\alpha} \exp\left[-\frac{\varepsilon_{\alpha}^{\mathrm{n}}}{k_B T}\right] \tag{5.5}$$

the nuclear partition function,

$$q_{\mathrm{e}} = \sum_{\beta} \exp\left[-\frac{\varepsilon_{\beta}^{\mathrm{e}}}{k_B T}\right] \tag{5.6}$$

the electronic partition function, and

$$q_{\mathrm{t}} = \sum_{\gamma} \exp\left[-\frac{\varepsilon_{\gamma}^{\mathrm{t}}}{k_B T}\right] \tag{5.7}$$

the translational partition function.

With this notation, the partition function q of an atom can be written as the product

$$q = q_{\mathrm{n}} q_{\mathrm{e}} q_{\mathrm{t}} \tag{5.8}$$

You see here at work the rule I mentioned earlier (Chapter 4, §3): if the total energy is a sum of energies, then the total partition function is a product of partition functions, each corresponding to one type of energy. In what follows I will show how to calculate q_{n}, q_{e}, and q_{t}.

The Translational Partition Function

§3. To evaluate q_{t} (given by Eq. 5.7), I need to know the energy eigenvalues $\varepsilon_i^{\mathrm{t}}$ of one particle in a box. Quantum mechanics gives, for a cubic box of side L,

$$\varepsilon_i^{\mathrm{t}} = \frac{h^2}{8mL^2}\left[n_x^2 + n_y^2 + n_z^2\right] \text{ with } n_x, n_y, n_z = 1, 2, \ldots, \infty \tag{5.9}$$

Here $h = 6.6262 \times 10^{-27}$ erg s is the "old" Planck constant (the "new" Planck constant is $\hbar = h/2\pi$) and m is the mass of the particle. The *quantum numbers* n_x, n_y, n_z are integers; each can take any value from 1 to infinity.

In the early manipulations, leading to Eq. 5.7, I labeled the single-particle state by an index γ. Now I find that the state of a particle in a box is characterized by three integers. For example, the state of lowest energy is $\{n_x, n_y, n_z\} = \{1, 1, 1\}$. Three states with higher energy are $\{n_x, n_y, n_z\} = \{2, 1, 1\}$, $\{n_x, n_y, n_z\} = \{1, 2, 1\}$, and $\{n_x, n_y, n_z\} = \{1, 1, 2\}$; they all have the same energy (see Eq. 5.9),

$$\frac{3h^2}{4mL^2} = \frac{h^2}{8mL^2}\left[2^2 + 1^2 + 1^2\right] = \frac{h^2}{8mL^2}\left[1^2 + 2^2 + 1^2\right] = \frac{h^2}{8mL^2}\left[1^2 + 1^2 + 2^2\right]$$

When we have several states with the same energy, we say that they are *degenerate*.

The fact that the state of the particle is described by three labels instead of one does not cause any trouble. Eq. 5.7 says: perform the sum over all the states of the particle. This means that I must sum over all values of the integers n_x, n_y, n_z. It follows that the partition function for translational motion is

$$\begin{aligned} q_{\text{t}} &= \sum_{n_x=1}^{\infty}\sum_{n_y=1}^{\infty}\sum_{n_z=1}^{\infty} \exp\left[-\frac{h^2}{8mL^2k_BT}\left(n_x^2 + n_y^2 + n_z^2\right)\right] \\ &= \left\{\sum_{n=1}^{\infty} \exp\left[-\frac{h^2n^2}{8mL^2k_BT}\right]\right\}^3 \end{aligned} \tag{5.10}$$

To get the last equality, I used $\exp[x+y] = \exp[x]\exp[y]$ and the fact that the sum over n_x is equal to the sum over n_y and to the sum over n_z.

§4. *The Sum in Eq. 5.10 Cannot be Evaluated by Brute Force.* In principle, I could perform the sum in Eq. 5.10 on a computer. Most often this is a splendid idea, but not now. For this particular sum a computer-aided brute-force evaluation is not possible; to converge the sum we would have to add too many terms.

Here is why. I can approximate the infinite sum in q_t by adding a finite number of terms, if the terms I ignore contribute very little to the value of the sum. In this particular case this means that once n satisfies

$$\exp\left[-\frac{h^2n^2}{8mL^2k_BT}\right] << 1 \tag{5.11}$$

I can stop adding the subsequent terms.

In Workbook SM5.1, I calculated the value of the expression $h^2/8mL^2k_BT$, which appears at the exponent in Eq. 5.11, for $V = 22$ liter, $m = 50$ amu, and $T = 300$ K.

I found that

$$\frac{h^2}{8mL^2k_BT} = 2.033 \times 10^{-21}$$

The condition in Eq. 5.11 therefore becomes

$$\exp\left[-2.033 \times 10^{-21}\, n^2\right] << 1$$

This is satisfied only when

$$n^2 \times 2.033 \times 10^{-21} \gg 1$$

To get a correct value for the sum defining q_t (Eq. 5.10), I have to add about 2.22×10^{10} terms. This is an exceedingly long calculation, even on a powerful computer.

Exercise 5.1

Write a program to calculate the sum in Eq. 5.10 and monitor how long it takes to add 1000 terms, 100,000 terms, etc.

---

§5. *Approximate the Sum by an Integral.* In many cases it is possible to approximate a sum by an integral. If the integral is easy to evaluate, this is a good method for calculating sums. I will use this method here to calculate the partition function q_t.

The connection between a sum of the form $\sum_{i=0}^{m} f(x_i)$ and the integral $\int_0^t f(x)dx$ is given by the following definition of the integral. Divide the x axis between 0 and t into equally spaced segments, delimited by the points $x_1, x_2, \ldots, x_m$ (see Fig. 5.1). The distance between any two neighboring points is

$$\Delta = x_{n+1} - x_n \tag{5.12}$$

The definition of the integral is (see any calculus book)

$$\int_0^t f(x)dx = \lim_{\Delta \to 0} \sum_{n=1}^{m} f(x_i)\Delta \tag{5.13}$$

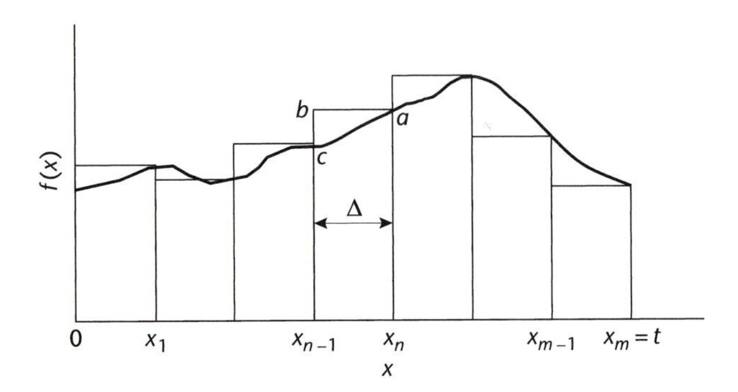

Figure 5.1 The rectangles approximate the area under the curve. The approximation gets better as Δ gets smaller.

The integral is equal to the sum when Δ is infinitesimally close to zero. If Δ is very small (but not arbitrarily close to zero) I can write

$$\sum_{n=1}^{m} f(x_i) \simeq \frac{\int_0^t f(x)dx}{\Delta} \qquad \text{if } \Delta \text{ is sufficiently small} \tag{5.14}$$

If I can calculate the integral in the right-hand side, this equation can be used to evaluate the sum. To do that it is necessary to find what Δ is for this particular example, and then show that it is sufficiently small so that the integral approximates well the sum. Before I do that I have to understand the meaning of the sentence "Δ is sufficiently small."

This understanding can be reached by using the theorem that states that the integral $\int_0^t f(x)dx$ is the area enclosed between the curve $f(x)$ and the x-axis. I will approximate this area (hence the integral) with the sum of certain rectangles built to cover the area approximately; this sum will turn out to be equal $\sum f(x_i)\Delta$. When the rectangles give a good approximation to the area under the curve, the integral is a good approximation to the sum. Here is how this idea is implemented.

On every segment (x_{n-1}, x_n), I construct a rectangle with the right-hand upper corner on the curve representing $f(x)$. One example is the rectangle defined by the points x_n, a, b, and x_{n-1} (Fig. 5.1). This rectangle has the area

$$\text{area}(n) = f(x_n)\Delta \tag{5.15}$$

The sum of all these areas

$$\sum_{n=1}^{m} f(x_n)\Delta \tag{5.16}$$

approximates the area under the curve (which is the integral). If $\Delta = x_{n+1} - x_n$ is large this approximation is poor. The error made can be easily evaluated. For example, for the rectangle defined by the points x_{n-1}, x_n, a, and b, the error is the area of the triangle defined by the points c, a, and b. This area is roughly equal to

$$\frac{(x_{n+1} - x_n)\bar{bc}}{2} = \frac{\Delta \bar{bc}}{2}, \tag{5.17}$$

where $\bar{bc}$ is the length of the segment joining b with c. As I put more points on the abscissa, the distance Δ between the points becomes smaller and so does the area of the triangles; the sum of the areas of the rectangles approximates better and better the area under the curve $f(x)$ (i.e. the integral). A reasonable approximation is obtained when the area of the triangles becomes much smaller than the area of the rectangles:

$$\frac{\bar{bc} \times \Delta}{2} \ll \text{area of rectangle defined by } x_{n-1}, x_n, a, \text{ and } b \tag{5.18}$$

From the figure it is clear that $\bar{bc} = f(x_n) - f(x_{n-1})$ and the area of the rectangle is $f(x_n)\Delta$. Using these two relationships in Eq. 5.18 leads to the condition

$$\frac{f(x_n) - f(x_{n-1})}{2} \ll f(x_n) \tag{5.19}$$

Now we know what we mean when we say that $\Delta \equiv x_n - x_{n-1}$ must be small: the neighboring points on the grid must be close enough to satisfy Eq. 5.19.

§6. *Apply this Analysis to Calculate the Partition Function* q_t. The sum in Eq. 5.10 is

$$\sigma \equiv \sum_{n=1}^{\infty} \exp\left[\frac{-h^2 n^2}{8mL^2 k_B T}\right] = \sum_{n=1}^{\infty} \exp\left[-x_n^2\right] \equiv \sum_{n=1}^{\infty} f(x_n) \tag{5.20}$$

I have used the following notation:

$$x_n^2 = \frac{h^2 n^2}{8mL^2 k_B T} \tag{5.21}$$

and

$$f(x) = e^{-x^2} \tag{5.22}$$

These manipulations have written σ in a form very similar to Eq. 5.14. This means that if Δ is sufficiently small we can calculate σ from (use Eq. 5.14)

$$\sigma \equiv \sum_{n=1}^{\infty} f(x_n) \approx \frac{\int_0^{\infty} f(x)dx}{\Delta} \tag{5.23}$$

This is not a mere repeat of Eq. 5.14: now we know that $f(x)$ is given by Eq. 5.22. To use Eq. 5.23 I must find what Δ is. This is easy to do: combine Eq. 5.12 defining Δ, with Eq. 5.21 defining x_n, to obtain

$$\Delta = x_{n+1} - x_n = (n+1)\sqrt{\frac{h^2}{8mL^2 k_B T}} - n\sqrt{\frac{h^2}{8mL^2 k_B T}} = \sqrt{\frac{h^2}{8mL^2 k_B T}} \equiv \Lambda(T) \tag{5.24}$$

The quantity

$$\Lambda(T) = \sqrt{\frac{h^2}{2\pi m k_B T}} \tag{5.25}$$

has units of length and is called the *de Broglie thermal wavelength*.

Our next step in evaluating σ (hence q_t) is to calculate the integral present in Eq. 5.23. This is easy to do by using **Mathematica** or **Mathcad** (see Workbook SM5.2):

$$\int_0^\infty f(x)dx = \int_0^\infty e^{-x^2} dx = \frac{\sqrt{\pi}}{2} \tag{5.26}$$

Introducing this in Eq. 5.23 and using the fact that $q_t = \sigma^3$ (combine Eq. 5.10 with the last equality in Eq. 5.20) leads to:

$$q_t = \left[\frac{L}{\Lambda(T)}\right]^3 = \frac{V}{\Lambda(T)^3} \tag{5.27}$$

where $V = L^3$ is the volume of the box (a cube of side L).

Exercise 5.2

Show that the de Broglie thermal wave length has units of length.

Exercise 5.3

Lord Rayleigh said that you are a mathematician if the equation $\int_0^\infty e^{-x^2} dx = \sqrt{\pi}/2$ is obvious to you. Try to derive this equation. Can you understand why π appears in it? *Hint*. Evaluate $\int_0^\infty e^{-y^2} dy \int_0^\infty e^{-z^2} dz = \left(\int_0^\infty e^{-x^2} dx\right)^2$.

§7. *Is this Evaluation of the Sum Correct?* To safely use this result we need to show that the condition given in Eq. 5.19 is satisfied. In Workbook SM5.2, I evaluated

$$f(y_n) - f(y_{n-1}) = \exp\left[-\frac{h^2 n^2}{8mL^2 k_B T}\right] - \exp\left[-\frac{h^2 (n-1)^2}{8mL^2 k_B T}\right]$$

for $n = 10$, $T = 298$ K, $m = 40$ au, and $L = 1$ cm (the volume is $V = L^3 = 1\ \text{cm}^3$), which is a typical situation. I found that it is practically equal to zero (see Workbook SM5.2), while

$$f(y_n) = \exp\left[-\frac{h^2 n^2}{8mL^2 k_B T}\right]$$

is nearly equal to 1. Clearly the condition in Eq. 5.19 is safely satisfied. For $n = 10^8$, a similar calculation leads to $3.93 \times 10^{-10} \ll 0.98$.

Now look what happens if the volume is very small and the particle is light. For an electron of mass 5.486×10^{-4} amu and a cube of side $L = 10^{-7}$ cm (that is, 10 Å), I have, for $n = 2$, $|f(y_{n-1}) - f(y_n)| = 4.41 \times 10^{-7}$ and $f(y_n) = 3.79 \times 10^{-26}$. The condition in Eq. 5.19 is not satisfied. We cannot approximate the sum with an integral. Is this a catastrophe? Not at all. For very small volumes and light particles, the exponentials in the original sum become negligible for rather small values of n. We can calculate the sum directly by adding a small number of terms.

Exercise 5.4

(a) Calculate q_t for an electron in a cubic box of side $L = 10$ Å by direct evaluation of the sum. Monitor the terms in the sum and find out for what value of n you can stop adding more terms. Stop when adding more terms will change the value of q_t by less than 0.1%.

(b) Do the same calculation for a proton and compare the two cases.

(c) Compare the results obtained by direct summation to those obtained when you approximate the sum by an integral.

§8. *Who Cares About Very Small Boxes?* So, the approximation of the sum by an integral fails when a particle is light and the box is very small (of order 10 Å = 1 nm). Who cares? Many people do. It is possible now to make small semiconductor or metal clusters of nanometer size. The electrons in these small crystals move almost freely. They don't have enough energy to get out of the crystal, which acts as a box. The thermodynamic properties of these electrons must be calculated by performing the sum in Eq. 5.10. The formula in Eq. 5.27, obtained by turning the sum into an integral, will not work. There is an additional complication that we do not discuss here. Unless the density of the mobile electrons is very low, Boltzmann statistics is not applicable.

Table 5.1 The nuclear spin I for several
indicates the mass of the nucleus. (From
and Physics, CRC Press, Boca Raton, Flori

Nucleus	I
H	$\frac{1}{2}$
^{3}He	$\frac{1}{2}$
^{4}He	0
^{16}O	0
^{17}O	$\frac{5}{2}$
^{18}O	$\frac{1}{2}$
^{19}F	$\frac{1}{2}$
^{19}Ne	$\frac{1}{2}$
^{35}Cl	$\frac{3}{2}$
^{37}Cl	$\frac{3}{2}$
^{40}Ar	0

Table 5.2 The degeneracy and electron
atoms. (From C. E. Moore, *Atomic Energ*
Circ. 1, 1949, p. 467.)

Atom	g_0^e	g_1^e	$\varepsilon_1^e - \varepsilon_0^e$ (eV)	g_2^e
H	2	2	10.20	–
He	1	3	19.82	–
O	5	3	0.02	1
F	4	2	0.05	6

As you can see, the values of $\varepsilon_1^e - \varepsilon_0^e$ for H
The same is true for Ar, Ne, Kr, Xe, and
electronic partition function is

$$q_e = g_0^e \exp$$

For other atoms, such as O, F, I, and Br, so
to $k_B T$, and for them we must use the full

Many chemists are interested in the properties of molecules contained in the small pores of specially prepared porous materials. In such cases, Eq. 5.27 must be applied with care, only if the condition in Eq. 5.19 is satisfied.

§9. *About Sums and Integrals.* We all spent quite a bit of time in school learning how to evaluate integrals, but we use that knowledge only when the integrals are particularly simple. In the past we used books containing tables with most integrals that can be evaluated analytically (for example, Gradshteyn and Ryzhik[a]). Recently we abandoned these tables in favor of computer languages (**Mathematica**, **Mathcad**) that perform symbolic manipulations. **Mathematica** managed to evaluate all the integrals in the tables and to find a number of errors in them. There are, however, many integrals that cannot be evaluated analytically. In such cases we resort to numerical methods which approximate the integrals by sums. Books have been written about the most efficient way of doing this. In this chapter we found ourselves going against the current: we approximated a sum by an integral. It is fairly rare that we have to do this. Keep this in mind: all integrals can be turned into sums but not all sums are integrals, because most of them do not satisfy Eq. 5.19.

The Nuclear Partition Function q_n

§10. *The Partition Function.* Let me write out explicitly the nuclear partition function. It is important to remember that the sum in any partition function is over all distinct states, not over all distinct energies. When you studied quantum mechanics, you learned that quantum states could be degenerate: several states can have the same energy. Let us see what happens in Eq. 5.5 if the ground state energy ε_0^n is triply degenerate (i.e. there are three states having the energy ε_0^n). The sum is

$$\begin{aligned} q_n &= \exp\left[-\frac{\varepsilon_0^n}{k_B T}\right] + \exp\left[-\frac{\varepsilon_0^n}{k_B T}\right] + \exp\left[-\frac{\varepsilon_0^n}{k_B T}\right] + \text{other terms} \\ &= 3\exp\left[-\frac{\varepsilon_0^n}{k_B T}\right] + \text{other terms} \end{aligned}$$

[a] I. S. Gradshteyn and I. M. Ryzhik, *Table of Integrals, Series, and Products* (translated by A. Jeffrey), Academic Press, New York, 1965.

In general, I can write

$$q_{\mathrm{n}} = \sum_{\mu}$$

where the sum is now over all distin
nuclear energy $\varepsilon_{\mu}^{\mathrm{n}}$ (i.e. the number of
Furthermore, I can factor $\exp\left[-\varepsilon_0^{\mathrm{n}}/k\right.$

$$q_{\mathrm{n}} = e^{-\varepsilon_0^{\mathrm{n}}/k_BT}\left(g_0^{\mathrm{n}} + g_1^{\mathrm{n}} \exp\left[-\frac{\varepsilon_1^{\mathrm{n}}}{k}\right.\right.$$

The difference $\varepsilon_i^{\mathrm{n}} - \varepsilon_0^{\mathrm{n}}$ is the excitatic
compared to k_BT. This means that all
The nuclear partition function is ther

$$q_{\mathrm{n}} = g$$

§11. *The Degeneracy of the Nuclea*
the nucleus is degenerate because of
degeneracy is

$$g_0^{\mathrm{n}}$$

Some values of the nuclear spin I are

The Electronic Partition Functic

§12. *The Partition Function.* The
the nuclear partition function and ca
of Eq. 5.29)

$$q_{\mathrm{e}} = \exp\left[-\frac{\varepsilon_0^{\mathrm{e}}}{k_BT}\right]\left(g_0^{\mathrm{e}} + g_1^{\mathrm{e}} \exp\right.$$

The degeneracy g_i^{e} of the electronic
the spin of the electrons. In Table 5.
atoms.

For example, the electronic partition function for the oxygen atom is

$$q_{\mathrm{e}} = \exp\left[-\frac{\varepsilon_0^{\mathrm{e}}}{k_BT}\right]\left(5 + 3 \times \exp\left[-\frac{0.02\ \mathrm{eV}}{k_BT}\right]\right.$$
$$\left. + 1 \times \exp\left[-\frac{0.03\ \mathrm{eV}}{k_BT}\right] + 5 \times \exp\left[-\frac{1.97\ \mathrm{eV}}{k_BT}\right]\right)$$

Since 1.97 eV $\gg k_BT$, I can stop at this term. The terms not included have higher energy and make an even smaller contribution to the sum.

The Choice of Zero Energy

§13. I am sure that you have heard many times that if you add a constant to the energy of the system, no measurable property changes. This is true in classical mechanics and quantum mechanics. As you will see, it is also true in statistical mechanics.

The partition function of the gas of atoms is (see Eqs 5.1 and 5.8)

$$Q = \left(\frac{qe}{N}\right)^N = \left(\frac{e}{N}\right)^N \left(q_{\mathrm{t}}q_{\mathrm{e}}q_{\mathrm{n}}\right)^N$$

If I use Eq. 5.27 for q_{t}, Eq. 5.30 for q_{n}, and Eq. 5.32 for q_{e} in this expression for Q, I obtain

$$Q = \left(\frac{e}{N}\right)^N \left(\frac{V}{\Lambda^3}\right)^N \exp\left[-\frac{N\varepsilon_0^{\mathrm{e}}}{k_BT}\right]\left(g_0^{\mathrm{e}} + \sum_{i=0}^{\infty} g_i^{\mathrm{e}} \exp\left[-\frac{\varepsilon_i^{\mathrm{e}} - \varepsilon_0^{\mathrm{e}}}{k_BT}\right]\right)^N$$
$$\left(g_0^{\mathrm{n}}\right)^N \exp\left[-\frac{N\varepsilon_0^{\mathrm{n}}}{k_BT}\right] \tag{5.34}$$

Since we can add any (positive or negative) constant to the energy, we can choose the energy scale so that

$$\varepsilon_0^{\mathrm{e}} + \varepsilon_0^{\mathrm{n}} = 0 \tag{5.35}$$

With this choice of energy, the partition function of the atom is

$$Q = \left(\frac{e}{N}\right)^N \left(\frac{V}{\Lambda^3}\right)^N \left(g_0^{\mathrm{n}}\right)^N \left(g_0^{\mathrm{e}} + g_1^{\mathrm{e}} \exp\left[-\frac{\varepsilon_1^{\mathrm{e}}}{k_BT}\right] + g_2^{\mathrm{e}} \exp\left[-\frac{\varepsilon_2^{\mathrm{e}}}{k_BT}\right] + \cdots\right)^N \tag{5.36}$$

In using this formula, one must sometimes remember that we used Eq. 5.35 to pick the zero of the energy scale. If, for example, you have a table in which the energies of the electronic states of an atom are $\mathcal{E} = \{\varepsilon_0^{\text{e}} = -14\text{ eV}, \varepsilon_1^{\text{e}} = -9\text{ eV}, \varepsilon_2^{\text{e}} = -4\text{ eV}, \ldots\}$, you cannot use these energies in Eq. 5.36. However, you can subtract from the values the constant -14 eV to obtain $\mathcal{E}' = \{\varepsilon_0^{\text{e}} = 0, \varepsilon_1^{\text{e}} = 5\text{ eV}, \varepsilon_2^{\text{e}} = 10\text{ eV}, \ldots\}$. Subtracting a constant from all the energies does not affect any measurable quantity, so the set $\mathcal{E}'$ is equivalent to $\mathcal{E}$.

However, you must be careful. If you use Eq. 5.36, you must use the set $\mathcal{E}'$, since it has $\varepsilon_0^{\text{e}} = 0$. If you want to use the set $\mathcal{E}$, you must use Eq. 5.34, which is not restricted to the cases in which $\varepsilon_0^{\text{e}} = 0$. Either calculation is correct: they lead to the same values for any measurable quantity. I will show you why this is so when I calculate the thermodynamic functions of the system.

§14. *Summary.* You now have everything you need for calculating the partition function of a gas of non-interacting atoms. This is given by

$$Q = \left(\frac{e}{N} q_{\text{t}} q_{\text{e}} q_{\text{n}}\right)^N$$

The translational partition function is

$$q_{\text{t}} = \frac{V}{\Lambda(T)^3}$$

where

$$\Lambda(T) = \frac{h^2}{\sqrt{2\pi m k_B T}}$$

is the de Broglie wavelength. The translational partition function depends on the volume of the gas, the mass of the atoms, temperature, and the universal constants h and k_B.

The electronic partition function is

$$q_{\text{e}} = \exp\left[-\frac{\varepsilon_0^{\text{e}}}{k_B T}\right]\left(g_0^{\text{e}} + g_1^{\text{e}} \exp\left[-\frac{\varepsilon_1^{\text{e}} - \varepsilon_0^{\text{e}}}{k_B T}\right] + g_2^{\text{e}} \exp\left[-\frac{\varepsilon_2^{\text{e}} - \varepsilon_0^{\text{e}}}{k_B T}\right] + \cdots\right)$$

Here ε_i^{e} are the energies of the electrons and g_i^{e} are their degeneracies. These are measured by spectroscopists or calculated by quantum mechanics. Often one ignores the factor $\exp\left[-\varepsilon_0^{\mathrm{e}}/k_B T\right]$, since this adds a constant to the free energy and such a constant affects no measurable quantity. For many atoms one can take

$$q_{\mathrm{e}} = g_0^{\mathrm{e}}.$$

The nuclear partition function is

$$q_{\mathrm{n}} = g_0^{\mathrm{n}}$$

where g_0^{n} is the degeneracy of the nuclear ground state. This exists because nuclei have $2I + 1$ states with the same energy. The values of I have been measured and tabulated.

Once we know the partition function of the gas, we can calculate all its thermodynamic properties. The next chapter shows how this is done.

6

THE THERMODYNAMIC FUNCTIONS OF AN IDEAL GAS OF ATOMS

Introduction

§1. In Chapter 1 I have postulated that the Helmholtz free energy A of a system is

$$A = -k_B T \ln Q \tag{6.1}$$

where Q is the partition function. From this, I can obtain all other thermodynamic functions by using Eqs 1.6–1.18. In this chapter, I implement this recipe for a gas of non-interacting atoms.

The partition function of a gas of non-interacting atoms is (see Chapter 5)

$$Q = \left(\frac{e}{N} q\right)^N = \left(\frac{e}{N} q_\text{t} q_\text{e} q_\text{n}\right)^N \tag{6.2}$$

with

$$q_\text{t} = \frac{V}{\Lambda(T)^3} \tag{6.3}$$

$$q_{\mathrm{n}} = g_0^{\mathrm{n}} \tag{6.4}$$

and

$$q_{\mathrm{e}} = g_0^{\mathrm{e}} + g_1^{\mathrm{e}} \exp\left[-\frac{\varepsilon_1^{\mathrm{e}} - \varepsilon_0^{\mathrm{e}}}{k_B T}\right] + g_2^{\mathrm{e}} \exp\left[-\frac{\varepsilon_2^{\mathrm{e}} - \varepsilon_0^{\mathrm{e}}}{k_B T}\right] + \cdots$$

$$= g_0^{\mathrm{e}} + \sum_{i=1}^{\infty} g_i^{\mathrm{e}} \exp\left[-\frac{\varepsilon_i^{\mathrm{e}} - \varepsilon_0^{\mathrm{e}}}{k_B T}\right] \tag{6.5}$$

Here

$$\Lambda(T) = \sqrt{\frac{h^2}{2\pi m k_B T}} \tag{6.6}$$

is the de Broglie thermal wavelength of a particle of mass m. V is the volume of the gas. g_0^{n} is the degeneracy of the ground state of the nucleus (see §10 on p. 67). g_i^{e} is the degeneracy of the i-th electronic state (see §12 on p. 68).

There is nothing subtle about the remainder of this chapter. We use the Helmholtz free energy A, defined by Eqs 6.1–6.6 in Eqs 1.6–1.18, which express all thermodynamic functions in terms of A. The implementation of this prescription is rather tedious and requires more patience than skill. When pertinent, I discuss the physical significance of the results.

Helmholtz Free Energy

§2. *Because I Divided Q by N!, A has the Correct Dependence on V and N.* Combining Eqs 6.1–6.5 gives

$$A = -k_B T \ln Q = -k_B T \ln \left(\frac{e}{N} q_{\mathrm{t}} q_{\mathrm{n}} q_{\mathrm{e}}\right)^N$$

$$= -N k_B T \ln \left(\frac{Ve}{N\,\Lambda(T)^3} q_{\mathrm{n}} q_{\mathrm{e}}\right) \tag{6.7}$$

The last equation was obtained by using Eq. 6.3 for q_{t}.

We know from thermodynamics that the free energy A is an *extensive* quantity. This means that if I keep temperature and pressure constant and double the number of moles, the free energy doubles. If Eq. 6.7 is correct, it must have this property. And indeed it does. Doubling the number of moles, at constant pressure and temperature, means that N changes to $2N$ and V changes to $2V$. The quantity V/N remains unchanged.

This is not surprising, since $V/N = V/nN_A = v/N_A$ where n is the number of moles, N_A is Avogadro's number (the number of atoms in a mole), and v is the molar volume. We know from thermodynamics that v is an *intensive* quantity: it depends on temperature and pressure but not on the amount of material in the system.

When I change N to $2N$ and V to $2V$, and keep T and p constant, the free energy change is

$$A \to -(2N)k_BT \ln\left(\frac{V}{N}\frac{eq_{\mathrm{n}}q_{\mathrm{e}}}{\Lambda(T)^3}\right) = 2A$$

This is the correct behavior, known from thermodynamics.

Why is it important to mention this? Remember that, when I derived (in Chapter 4) the partition function for a system of independent particles, I decided to divide the result by $N!$. In the early days of statistical mechanics, this was a patch that made the theory work. The presence of $N!$ in Q introduced the factor e/N in Eq. 6.2. Without that factor, the free energy would have been

$$A = -N\,k_BT \ln\left(\frac{Vq_{\mathrm{n}}q_{\mathrm{e}}}{\Lambda^3}\right) \tag{6.8}$$

Doubling V and N in this equation, while keeping T and p fixed, will make A change to $-2Nk_BT\ln\left(2Vq_{\mathrm{n}}q_{\mathrm{e}}/\Lambda^3\right)$, not $2A$. Thus if I use Eq. 6.8, A does not change to $2A$ as it should. Eq. 6.8, obtained without division by $N!$, is fatally flawed. $N!$ was originally introduced in the theory to fix this flaw. Later, quantum mechanics was used to derive the result rigorously.

§3. *The Free Energy is a sum of a Collective, a Translational, an Electronic, and a Nuclear Contribution.* Eq. 6.1, and the properties of the logarithm, allow me to write the Helmholtz free energy as

$$A = A_{\mathrm{t}} + A_{\mathrm{e}} + A_{\mathrm{n}} + A_{\mathrm{c}} \tag{6.9}$$

with

$$A_t \equiv -Nk_BT \ln q_t = -Nk_BT \ln\left(\frac{V}{\Lambda(T)^3}\right) \tag{6.10}$$

$$A_e \equiv -Nk_BT \ln\left(q_e\right) \tag{6.11}$$

$$A_n \equiv -Nk_BT \ln\left(q_n\right) = -Nk_BT \ln\left(g_0^n\right) \tag{6.12}$$

$$A_c = -k_BTN \ln\left(\frac{e}{N}\right) \tag{6.13}$$

The free energies A_t, A_e, and A_n originate from the translational, electronic, and nuclear degrees of freedom. The *collective* free energy A_c comes from the division of the partition function with $N!$. It is called 'collective' because it is not a property of an atom, but of a collection of identical particles. Sometimes A_t and A_c are combined and the sum $A_t + A_c$ is called the translational Helmholtz free energy.

When we do not want to distinguish separate contributions, the Helmholtz free energy of the gas is

$$A = -Nk_BT\left(\ln\frac{Ve}{N\Lambda(T)^3} + \ln q_e + \ln g_0^n\right) \tag{6.14}$$

Since all thermodynamic quantities are either derivatives of A or simple expressions involving A, they will all be sums of translational, electronic, nuclear, and collective contributions.

In what follows, I will use Eqs 6.9–6.13 to derive expressions for the energy U, the entropy S, the pressure p, the heat capacity at constant pressure C_p, the heat capacity at constant volume C_v, and the chemical potential μ.

The Energy *U*

§4. From the Helmholtz free energy, I can derive a formula for the energy by using Eq. 1.9, which is

$$U = -T^2\frac{\partial}{\partial T}\left(\frac{A}{T}\right)_{N,V} \tag{6.15}$$

From this, and the fact that $A = A_t + A_e + A_n + A_c$, I have

$$U = U_t + U_e + U_n + U_c \tag{6.16}$$

with

$$U_{\mathrm{t}} = Nk_BT^2\frac{\partial \ln q_{\mathrm{t}}}{\partial T} = Nk_BT^2\frac{\partial}{\partial T}\left[\ln\left(\frac{V}{\Lambda(T)^3}\right)\right] \tag{6.17}$$

$$U_{\mathrm{e}} = Nk_BT^2\frac{\partial \ln q_{\mathrm{e}}}{\partial T} \tag{6.18}$$

$$U_{\mathrm{n}} = Nk_BT^2\frac{\partial \ln q_{\mathrm{n}}}{\partial T} \tag{6.19}$$

$$U_{\mathrm{c}} = Nk_BT^2\frac{\partial \ln q_{\mathrm{c}}}{\partial T} \tag{6.20}$$

There is no nuclear or collective contribution to internal energy, because A_{n}/T and A_{c}/T are independent of temperature and the derivatives in Eqs 6.19 and 6.20 are zero.

§5. *The Translational Contribution U_{t} to Energy.* Using properties of the logarithm function ($\ln(ab) = \ln a + \ln b$ and $\ln a^b = b\ln a$) to rewrite Eq. 6.17 gives

$$\begin{aligned} U_{\mathrm{t}} &= Nk_BT^2\frac{\partial}{\partial T}\left(\ln\frac{V}{\Lambda(T)^3}\right)_{N,V} \\ &= Nk_BT^2\frac{\partial}{\partial T}\left[\ln(V) - 3\ln\left(\Lambda(T)\right)\right]_{N,V} \end{aligned}$$

The only temperature-dependent quantity under the partial derivative is $\Lambda(T)$, so

$$U_{\mathrm{t}} = -3Nk_BT^2\frac{\partial}{\partial T}\left(\ln\Lambda(T)\right)_{N,V} \tag{6.21}$$

Since $\Lambda(T) = \sqrt{h^2/2\pi mk_BT}$, I have $\ln\Lambda(T) = -\frac{1}{2}\ln T + \frac{1}{2}\ln(h^2/2\pi mk_B)$. Taking the derivative gives

$$\frac{\partial \ln\Lambda(T)}{\partial T} = -\frac{1}{2T}; \tag{6.22}$$

using this in Eq. 6.21 leads to

$$U_{\mathrm{t}} = \frac{3}{2}Nk_BT \tag{6.23}$$

The same result is obtained by using a computer in Workbook SM.6.

§6. *The Mean Velocity.* Since the atoms in the gas have only kinetic energy (there is no interaction energy), U_t must be the mean kinetic energy of N molecules. Dividing by N or taking $N = 1$ gives the mean translational energy per molecule:

$$u_t \equiv \frac{U_t}{N} = \frac{3}{2}k_B T \tag{6.24}$$

Each molecule contributes an amount $3k_BT/2$ to the total energy.

Since the mean kinetic energy is the average of $mv^2/2$, I have

$$\frac{m\langle v^2\rangle}{2} = \frac{3}{2}k_B T$$

Here $\langle v^2\rangle$ denotes the average of v^2. From this equation I can estimate how fast a molecule moves through the gas:

$$\sqrt{\langle v^2\rangle} = \sqrt{\frac{3k_BT}{m}}$$

For example, for Ne, I obtain (use CGS units: $k_B = 1.3806 \times 10^{-16}$ erg/K, $m = 20 \times 1.6605 \times 10^{-24}$ g)

$$\sqrt{\langle v^2\rangle} = \sqrt{\frac{3 \times 1.3806 \times 10^{-16} \times 300}{20 \times 1.6605 \times 10^{-24}}}$$

$$= 61{,}167.5 \text{ cm/s} \ = 2{,}202 \text{ km/h}$$

This is a lot faster than an airplane.

Exercise 6.1

Your skin is bombarded by molecules traveling at 2202 km/h. Do they break bonds in the molecules forming your skin? Explain how you reached your conclusion.

---

§7. *Joule's Experiments.* Eq. 6.23 is in agreement with Joule's experiments which have shown that the energy of an ideal gas does not depend on volume. This is easy to understand: if we change the volume, we change the mean distance between the molecules; if the molecules do not interact, this change of distance causes no change in the energy.

§8. *The Electronic Contribution* U_e. The electronic contribution U_e is more complicated. I have (see Eq. 6.18)

$$U_e = N k_B T^2 \frac{\partial}{\partial T} \left(\ln q_e \right)_{N,V} = N k_B T^2 \frac{1}{q_e} \left(\frac{\partial q_e}{\partial T} \right)_{N,V} \tag{6.25}$$

The electronic partition function is (see Eq. 6.5)

$$q_e = g_0^e + g_1^e \exp\left[-\frac{\varepsilon_1^e - \varepsilon_0^e}{k_B T} \right] + g_2^e \exp\left[-\frac{\varepsilon_2^e - \varepsilon_0^e}{k_B T} \right] + \cdots \tag{6.26}$$

Inserting this into Eq. 6.25 and taking the derivatives (see Workbook SM.6) gives

$$U_e = \frac{N}{q_e} \sum_{i=1}^{\infty} (\varepsilon_i^e - \varepsilon_0^e) g_i^e \exp\left[-\frac{\varepsilon_i^e - \varepsilon_0^e}{k_B T} \right] \tag{6.27}$$

The terms in the sum Eq. 6.27 become smaller and smaller as i increases. We evaluate the sum by adding terms until the addition of a new term changes U_e by a small amount.

While I used **Mathematica** to derive Eq. 6.27 from Eq. 6.26, it would be useful practice to derive it long-hand.

Eq. 6.27 is a very natural result. The quantity

$$p_i = \frac{1}{q_e} g_i^e \exp\left[-\frac{\varepsilon_i^e}{k_B T} \right]$$

is the probability that the atom has the electronic energy ε_i^e. With this notation, Eq. 6.27 becomes

$$U_e = N \sum_{i=1}^{\infty} \varepsilon_i^e p_i$$

This is just the mean electronic energy of the atom, multiplied with the number of atoms.

§9. *The Total Energy.* To obtain the total energy of the gas, we add the translational and the electronic contributions (the nuclear and the collective contributions are zero) and obtain:

$$U = \frac{3}{2}k_B T + \frac{1}{q_e}\sum_{i=1}^{\infty} \varepsilon_i^e g_i^e \exp\left[-\frac{\varepsilon_i^e - \varepsilon_0^e}{k_B T}\right] \tag{6.28}$$

For many atoms, $\varepsilon_i^e - \varepsilon_0^e \gg k_B T$ when $i \geq 1$, and the electronic degrees of freedom do not contribute to energy. In those cases,

$$U = \frac{3}{2}k_B T \tag{6.29}$$

Pressure

§10. The pressure is given by Eq. 1.7:

$$p = -\left(\frac{\partial A}{\partial V}\right)_{T,N}$$

Using Eqs 6.9–6.13 for A, and noting that A_e, A_n, and A_c do not depend on volume, gives

$$p = -\left(\frac{\partial A_t}{\partial V}\right)_{T,N} = -\frac{\partial}{\partial V}\left[-k_B T \ln\frac{V}{\Lambda^3}\right]_{N,T}$$

$$= k_B T N \frac{\partial}{\partial V}\left(\ln V + \ln\frac{1}{\Lambda^3}\right)_{N,T} = \frac{k_B T N}{V}$$

Since $N = nN_A$ where n is the number of moles and N_A is Avogadro's number, I can write this equation as

$$p = \frac{nk_B N_A T}{V} \tag{6.30}$$

If I identify $k_B N_A \equiv R$, Eq. 6.30 is the ideal gas equation of state $pV = nRT$, obtained experimentally. Since R is known from measurements, this equation allows us to determine the value of k_B.

The fact that only the translational free energy contributes to the pressure of the gas accords with our intuition that the pressure is due to the collision of the molecules with the walls.

Exercise 6.2

Use the equations derived so far to find a connection between the energy U and the pressure.

Entropy

§11. I can derive a formula for the entropy S of N atoms from Eq. 1.6:

$$S = -\left(\frac{\partial A}{\partial T}\right)_{N,V} \tag{6.31}$$

By using Eqs 6.9–6.13, I can rewrite this as

$$S = S_{\mathrm{t}} + S_{\mathrm{e}} + S_{\mathrm{n}} + S_{\mathrm{c}} \tag{6.32}$$

with

$$S_{\mathrm{t}} = -\frac{\partial}{\partial T}\left[-k_B T N \ln\left(\frac{V}{\Lambda(T)^3}\right)\right]_{V,N} \tag{6.33}$$

$$S_{\mathrm{e}} = -\frac{\partial}{\partial T}\left[-k_B T N \ln q_{\mathrm{e}}(T)\right]_{V,N} \tag{6.34}$$

$$S_{\mathrm{n}} = -\frac{\partial}{\partial T}\left[-k_B T N \ln q_{\mathrm{n}}\right]_{V,N} \tag{6.35}$$

$$S_c = -\frac{\partial}{\partial T}\left[-k_B T N \ln\left(\frac{N}{e}\right)\right]_{V,N} \tag{6.36}$$

Note that entropy is the first thermodynamic function to be affected by the collective and the nuclear partition functions.

§12. *Nuclear Contribution to Entropy.* The nuclear contribution S_{n} is easy to calculate: $q_{\mathrm{n}} = g_0^{\mathrm{n}}$ is independent of temperature and therefore

$$S_{\mathrm{n}} = k_B N \ln g_0^{\mathrm{n}} \tag{6.37}$$

If the ground state is not degenerate, then $g_0^{\mathrm{n}} = 1$ and $S_{\mathrm{n}} = 0$.

For 1 mole of substance, $N = N_A$ and $k_B N_A = R$. Therefore the nuclear contribution to entropy per mole is $R \ln g_0^{\mathrm{n}}$. Since $R = 1.98$ cal/mol K and the entropy of a gas is of order of tens of cal/mol K, this is a sizeable fraction of the total entropy. However, in most cases, when we calculate observable quantities, we need entropy differences, such as $\Delta S \equiv S(T,p=1) - S(298,p=1)$. Since changing the temperature has no effect on g_0^n, the contribution $R \ln g_0^n$ cancels out when ΔS is calculated.

In Chapter 3 we discussed the Third Law of thermodynamics, which states that entropy goes to zero when temperature goes to zero, but then also mentioned that there are exceptions from this rule. One such exception comes from the nuclear degeneracy, which is in turn due to the nuclear spin I. Since $g_0^n = 2I + 1$, the nuclear contribution to entropy is zero only if the nucleus of the atom has spin $I = 0$.

Exercise 6.3

In Chapter 3 we connected entropy to information by analyzing the equation $S = -k_B \sum_{i=0}^{n} p_i \ln(p_i)$. Use this equation, and the assumption that all spin states of the nucleus are equally probable, to derive Eq. 6.37.

§13. *Translational Contribution to Entropy.* The translational contribution to entropy is calculated next. Using Eq. 6.33 for S_t gives

$$S_t = -\frac{\partial}{\partial T}\left[-k_B T N \ln\left(\frac{V}{\Lambda(T)^3}\right)\right]_{V,N}$$

$$= k_B N \ln\left[\frac{V}{\Lambda(T)^3}\right] + k_B N T \frac{\partial}{\partial T}\left[\ln(V) - 3\ln\Lambda(T)\right]_{V,N}$$

$$= k_B N \ln\left(\frac{V}{\Lambda(T)^3}\right) - 3k_B N T \frac{\partial \ln \Lambda(T)}{\partial T}$$

$$= k_B N \ln \frac{V}{\Lambda(T)^3} + \frac{3}{2} k_B N \tag{6.38}$$

To obtain the last equality, I used Eq. 6.22 for the derivative of $\ln \Lambda$ with respect to T.

Note that the experiments tell us that the total entropy is *an extensive quantity*: if we double the volume and the number of moles, but keep the pressure and the temperature constant, the entropy doubles (note that doubling V and N leaves the molar volume $v = V/N$ unchanged). It is straightforward to verify that the translational entropy given by Eq. 6.38 does not have this property. This is a grave concern, but not a fatal one: we cannot measure the translational entropy, only the total entropy. We can still hope that the total entropy is extensive.

To calculate the translational entropy per mole, replace N with N_A and use $k_B N_A = R$. This gives

$$S_{\mathrm{t}} = R\left(\frac{3}{2} + \ln\frac{V}{\Lambda^3}\right) \quad \text{(per mole)} \tag{6.39}$$

§14. *Electronic Contribution to Entropy.* The electronic contribution is more complicated. From Eq. 6.34, I obtain

$$\begin{aligned} S_{\mathrm{e}} &= -\frac{\partial}{\partial T} k_B T N \ln q_{\mathrm{e}} \\ &= k_B N \ln q_{\mathrm{e}}(T) + \frac{k_B T N}{q_{\mathrm{e}}(T)} \frac{\partial q_{\mathrm{e}}(T)}{\partial T} \\ &= k_B N \ln q_{\mathrm{e}}(T) + \frac{k_B N}{q_{\mathrm{e}}(T)} \sum_{i=1}^{\infty} \frac{g_i^{\mathrm{e}}(\varepsilon_i^{\mathrm{e}} - \varepsilon_0^{\mathrm{e}})}{k_B T} \exp\left[-\frac{\varepsilon_i^{\mathrm{e}} - \varepsilon_0^{\mathrm{e}}}{k_B T}\right] \end{aligned} \tag{6.40}$$

For the second equality, I used Eq. 6.26 and simple calculus (see also Workbook SM.6).

If you compare Eq. 6.40 with Eq. 6.27, you will notice that you can write

$$S_{\mathrm{e}} = k_B N \ln q_{\mathrm{e}}(T) + \frac{U_{\mathrm{e}}(T)}{T} \tag{6.41}$$

Either Eq. 6.40 or Eq. 6.41 can be used in numerical calculations.

Exercise 6.4

Show that S_{e} and S_{n} are extensive quantities.

§15. *Collective Contribution to Entropy.* This is given by

$$\begin{aligned} S_c &= -\frac{\partial}{\partial T}\left[k_B T N \ln\left(q_c\right)\right] \\ &= -\frac{\partial}{\partial T}\left[k_B T N \ln\left(\frac{e}{N}\right)\right] \\ &= k_B N - k_B N \ln\left(N\right) \end{aligned} \tag{6.42}$$

Above I used $\ln(e) = 1$.

Exercise 6.5

Show that S_c is not an extensive quantity.

§16. *The Total Entropy.* The total entropy is

$$\begin{aligned} \frac{S}{k_B N} &= \frac{S_t + S_e + S_n + S_c}{k_B N} \\ &= \ln\frac{V}{N\Lambda(T)^3} + \frac{5}{2} + \ln g_0^n + \ln q_e(T) \\ &+ \frac{1}{q_e}\sum_{i=1}^{\infty}\frac{g_i^e \varepsilon_i^e}{k_B T}\exp\left[-\frac{\varepsilon_i^e - \varepsilon_0^e}{k_B T}\right] \end{aligned} \tag{6.43}$$

If $\varepsilon_1^e - \varepsilon_0^e \gg k_B T$, the last two terms can be neglected.

It is easy to verify that the total entropy in Eq. 6.43 *is an extensive quantity.* This happens because the sum $S_t + S_c$ is extensive, although the individual terms are not.

Let us emphasize again the beneficial role of the term $N!$, which generates the collective contribution to the partitions functions. Without it the total entropy would not be extensive and it will also disagree with the experiments. It was the analysis of the entropy of noble gases that led Sachur and Tetrode to introduce the term $N!$ in the partition function, to bring the computed entropy in agreement with experiments.

Exercise 6.6

Calculate S from the relationship $A = U - TS$, by using Eqs 6.9–6.13 for A and Eqs 6.23 and 6.27 for U.

The Heat Capacities C_v and C_p

§17. The heat capacity at constant volume C_v can be calculated from Eq. 1.11:

$$C_v = \left(\frac{\partial u}{\partial T}\right)_{V,N} \tag{6.44}$$

Heat capacity is defined for 1 mole of substance and this is why I use the energy u of 1 mole of substance (not U) in the right-hand side of Eq. 6.44.

Once C_v is calculated from Eq. 6.44, it is easy to obtain the heat capacity at constant pressure C_p from

$$C_p = C_v + R \tag{6.45}$$

This equation was derived in thermodynamics and it is valid for ideal gases only.

To evaluate C_v, I use the results derived in §4 on p. 76 for the energy U. To calculate the energy u (per mole) I replace N with N_A in the equations for U, and also use $k_B N_A = R$.

It is easier to follow the derivation if I collect here the equations for u that I need. From Eq. 6.23 the translational energy of a mole of gas is

$$u_t = \frac{3}{2}RT \tag{6.46}$$

Eq. 6.27 gives the electronic contribution:

$$u_e = \frac{N_A}{q_e(T)} \sum_{i \geq 1} \varepsilon_i^e g_i^e \exp\left[-\frac{\varepsilon_i^e}{k_B T}\right] \tag{6.47}$$

The electronic partition function is (see Eq. 6.26):

$$q_e = g_0^e + g_1^e \exp\left[-\frac{\varepsilon_1^e - \varepsilon_0^e}{k_B T}\right] + g_2^e \exp\left[-\frac{\varepsilon_2^e - \varepsilon_0^e}{k_B T}\right] + \cdots \tag{6.48}$$

Only the translational and the electronic degrees of freedom affect energy (see §5 on p. 77) and therefore

$$u = u_{\mathrm{t}} + u_{\mathrm{e}} \tag{6.49}$$

Using this in Eq. 6.44 gives

$$C_v = (C_v)_{\mathrm{t}} + (C_v)_{\mathrm{e}} \tag{6.50}$$

with

$$(C_v)_{\mathrm{t}} = \left(\frac{\partial u_{\mathrm{t}}}{\partial T}\right)_{V,N} \tag{6.51}$$

$$(C_v)_{\mathrm{e}} = \left(\frac{\partial u_{\mathrm{e}}}{\partial T}\right)_{V,N} \tag{6.52}$$

After this preparation we are now ready to derive equations for C_v and C_p.

§18. *The Translational Contribution.* We have

$$(C_v)_{\mathrm{t}} = \frac{\partial}{\partial T}(u_{\mathrm{t}})_{N,V} = \frac{\partial}{\partial T}\left(\frac{3RT}{2}\right)_{N,V} = \frac{3R}{2} \tag{6.53}$$

This, together with Eq. 6.45, gives

$$(C_p)_{\mathrm{t}} = (C_v)_{\mathrm{t}} + R = \frac{5}{2}R \tag{6.54}$$

Exercise 6.7

Use the relationship $C_v/T = (\partial s/\partial T)_{V,N}$, where s is the entropy of 1 mole of gas, to derive an expression for $(C_v)_{\mathrm{t}}$. Does it agree with the one derived above?

§19. *The Electronic Contribution to C_v.* We have

$$(C_v)_{\mathrm{e}} = \left(\frac{\partial u_{\mathrm{e}}}{\partial T}\right)_{V,N}$$

$$= N_{\mathrm{A}}\frac{\partial}{\partial T}\left(\frac{\sum_{i\geq 1}\left(\varepsilon_i^{\mathrm{e}} - \varepsilon_0^{\mathrm{e}}\right) g_i^{\mathrm{e}} \exp\left[-\left(\varepsilon_i^{\mathrm{e}} - \varepsilon_0^{\mathrm{e}}\right)/k_B T\right]}{q_{\mathrm{e}}(T)}\right)_{V,N}$$

The evaluation of this derivative is tedious, because q_e is a function of temperature. I performed it in Workbook SM.6, but with a little patience it can be evaluated "by hand". The result is

$$(C_v)_e = k_B N_A \left\{ \frac{1}{q_e(T)} \sum_{i \geq 1} \left(\frac{\varepsilon_i^e - \varepsilon_0^e}{k_B T} \right)^2 g_i^e \exp\left[-\frac{\varepsilon_i^e - \varepsilon_0^e}{k_B T} \right] - \frac{1}{q_e(T)^2} \left(\sum_{i \geq 1} g_i^e \left(\frac{\varepsilon_i^e - \varepsilon_0^e}{k_B T} \right) \exp\left[-\frac{\varepsilon_i^e - \varepsilon_0^e}{k_B T} \right] \right)^2 \right\} \quad (6.55)$$

The right-hand side of Eq. 6.55 is easy to evaluate because $\left(\varepsilon_i^e - \varepsilon_0^e\right)/k_B T$ grows rapidly with i, and therefore the terms $\exp\left[-\left(\epsilon_i - \epsilon_0\right)/k_B T\right]$ decay rapidly with i. As a result, we need only a few terms in the sum.

§20. *Comments Regarding C_v and C_p.* The theory makes some striking predictions. As long as $\varepsilon_i^e - \varepsilon_0^e \gg k_B T$, the heat capacity C_v at constant volume of *all atomic gases* is equal to $\frac{3}{2}R$ and $C_p = \frac{5}{2}R$. This prediction is applicable to ideal gases only (low density, which means low pressure and/or high temperature). The thermodynamic data for the noble gases (He, Ne, Ar, Kr, and Xe) in the NIST database (webbook.nist.gov/chemistry) show that between 298.15 K and 6000 K, the values of C_p do not change with temperature and are equal to $\frac{5}{2}R$. These data are at $p = 1$ atm, where the noble gases are ideal.

The predictions of the present theory are not valid when a gas is no longer ideal. For example, at 10 atm, C_p for argon is 21.42 J/mol K at 250 K, 21.15 J/mol K at 320 K, and 20.97 J/mol K at 600 K. The gas is not ideal at $p = 10$ atm and C_p starts to depend on temperature. Furthermore, the heat capacity also depends on pressure: at 260 K, $C_p = 21.42$ J/mol K when $p = 10$ atm, 22.14 J/mol K when $p = 20$ atm, and 36.76 J/mol K when $p = 200$ atm.

The Chemical Potential

§21. I can calculate the chemical potential from Eq. 1.16:

$$\mu = \left(\frac{\partial A}{\partial N} \right)_{T,V} \quad (6.56)$$

Since $A = A_t + A_e + A_n + A_c$, the chemical potential splits into a sum of four terms:

$$\mu = \mu_t + \mu_e + \mu_n + \mu_c \tag{6.57}$$

§22. *The Translational Contribution to* μ. The translational contribution is

$$\begin{aligned}\mu_t &= \left(\frac{\partial A_t}{\partial N}\right)_{T,V} = \frac{\partial}{\partial N}\left(-k_B T N \ln \frac{V}{\Lambda(T)^3}\right) \\ &= -k_B T \ln \frac{V}{\Lambda(T)^3}\end{aligned} \tag{6.58}$$

This is the chemical potential per molecule. To obtain the chemical potential per mole I multiply this equation by Avogadro's number N_A and use $R = k_B N_A$. The result is

$$\mu_t = -RT \ln \frac{V}{\Lambda(T)^3} \tag{6.59}$$

Note that μ must be an extensive quantity and that μ_t given by Eq. 6.59 is not. Given the experience gained when we calculated the entropy, we are not going to panic. We expect that the collective contribution to μ will combine with the translational one, to repair this flaw.

§23. *The Nuclear Contribution to* μ. The nuclear contribution is (use Eq. 6.12)

$$\mu_n = \left(\frac{\partial A_n}{\partial N}\right)_{T,V} = \frac{\partial}{\partial N}\left(-k_B T N \ln g_0^n\right) = -k_B T \ln g_0^n \tag{6.60}$$

This is the chemical potential per molecule. The value per mole is

$$\mu_n = -RT \ln g_0^n \tag{6.61}$$

§24. *The Electronic Contribution to* μ. The electronic contribution is also simple, because q_e does not depend on N. It is given by

$$\mu_e = \left(\frac{\partial A_e}{\partial N}\right)_{T,V} = \frac{\partial}{\partial N}\left(-k_B T N \ln q_e\right) = -k_B T \ln q_e(T) \tag{6.62}$$

This is the electronic contribution per molecule. The value per mole is

$$\mu_e = -RT \ln q_e(T) \tag{6.63}$$

The electronic partition function $q_e(T)$ is given by Eq. 6.5.

§25. *The Collective Contribution* μ_c. Using the collective contribution to the Helmholtz free energy (Eq. 6.13), we calculate the collective contribution μ_c to be

$$\begin{aligned} \mu_c &= \left(\frac{\partial A_c}{\partial N}\right)_{T,V} \\ &= \frac{\partial}{\partial N}\left[-k_B TN \ln\left(\frac{e}{N}\right)\right]_{T,V} \\ &= -k_B T \ln(N) \end{aligned} \tag{6.64}$$

Since we have calculated $(\partial A_c/\partial N)_{T,V}$ we have obtained the chemical potential per molecule. To get the chemical potential per mole, which is normally used in thermodynamics, we multiply the above expression with the Avogadro number N_A, and replace N with N_A. Thus we have

$$\mu_c = RT \ln(N_A) \tag{6.65}$$

Like the translational contribution μ_t, the collective contribution μ_c is not an extensive quantity.

For future reference I write down the sum of the translational and collective contributions to the chemical potential (per mole)

$$\mu_t + \mu_c = -RT \ln \frac{V}{\Lambda(T)^3} + RT \ln(N_A) = -RT \ln\left(\frac{V}{N_A \Lambda(T)^3}\right) \tag{6.66}$$

§26. *The Total Chemical Potential.* According to Eq. 6.57, adding the four separate contributions, μ_t, μ_n, μ_e, and μ_c, gives the total chemical potential. Using Eqs 6.59, 6.61, 6.63, and 6.65, for the separate contributions to μ, gives the following expression for the chemical potential of a mole of gas:

$$\mu = -RT\left(\ln \frac{V}{N_A \Lambda(T)^3} + \ln g_0^n + \ln q_e(T)\right) \tag{6.67}$$

As expected, the translational and collective contributions have combined (see Eq. 6.66) in the expression for the chemical potential of the gas, to give the term $v \equiv V/N_A$, which is the molar volume of the gas. Since v is an extensive quantity, μ ends up being extensive, as it must be.

§27. *Chemical Potential as a Function of Pressure.* It is customary in thermodynamics to write the chemical potential in the form

$$\mu(T,p) = \mu_0(T) + RT \ln(p/p_0) \tag{6.68}$$

where $p_0 = 1$ atm (or 1 bar, depending on the definition of the standard state). Here I will write Eq. 6.67 in the form Eq. 6.68 and derive an expression for $\mu_0(T)$. I replace V in Eq. 6.67 with the expression RT/p, given by the ideal gas law, and manipulate the result by using the properties of the logarithm ($\ln(ab) = \ln a + \ln b$ and $-\ln(a/b) = \ln(b/a)$):

$$\begin{aligned} \mu &= -RT \ln\left(\frac{RT}{N_A p}\frac{p_0}{p_0}\frac{1}{\Lambda^3}\right) - RT \ln\left(g_0^n q_e(T)\right) \\ &= -RT \ln\left(\frac{RT g_0^n q_e(T)}{p_0 N_A \Lambda^3}\right) + RT \ln \frac{p}{p_0} \end{aligned} \tag{6.69}$$

Compare this to Eq. 6.68 and you see that

$$\mu_0(T) = -RT \ln\left(\frac{RT}{p_0 N_A \Lambda(T)^3} g_0^n q_e(T)\right) \tag{6.70}$$

The chemical potential per molecule is (divide Eq. 6.69 by N_A, and use $k_B = R/N_A$)

$$\mu(T,p) = -k_B T \ln\left(\frac{RT}{p_0 N_A \Lambda(T)^3} g_0^n q_e(T)\right) + k_B T \ln\left(\frac{p}{p_0}\right) \tag{6.71}$$

Note that $\Lambda(T)^3$ is the volume of cube having sides equal to the de Broglie wavelength of the atom and $RT/(p_0 N_A)$ is the volume per molecule, when the pressure is 1 atm. $RT/(p_0 N_A \Lambda^3)$ is a dimensionless number. In an ideal gas this number is greater than 1 and therefore μ_0 is negative.

Exercise 6.8

Using the dependence of Λ on mass, find the relationship among the standard chemical potentials μ_0 of H, He, Ar, Kr, and Xe.

Exercise 6.9

Show that you can write $\mu_0(T)$ in the form

$$-RT \ln\left(\frac{V_0}{N_A \Lambda(T)^3} g_0^{\mathrm{n}} q_{\mathrm{e}}(T)\right)$$

where V_0 is the volume of 1 mole of ideal gas at temperature T. Calculate $N_A \Lambda(T)^3$ for several temperatures for He and Xe, and compare to V_0. Can you find a temperature at which $V_0 = \Lambda(T)^3$? What is the value of μ_0 at this temperature if $\varepsilon_1^{\mathrm{e}} - \varepsilon_0^{\mathrm{e}} \gg k_B T$ and $g_0^{\mathrm{n}} = 1$?

Summary

There is either too much or too little to summarize. I started with the recipes for calculating the Helmholtz free energy A (given in Chapter 1) from the partition function Q of the gas, calculated in Chapter 4. Once A was determined, I calculated the other thermodynamic quantities by using the expressions given in Chapter 1.

The rest of the chapter consisted in a patient application of these formulae to obtain expressions for the energy, the pressure, the entropy, the heat capacity, and the chemical potential of an ideal gas of atoms.

Occasionally, I showed that some striking qualitative predictions contained in these equations are in agreement with experiment. In the next two chapters, we compare the values calculated from these equations with the results obtained by measurement.

7

THE THERMODYNAMIC PROPERTIES OF AN IDEAL GAS FOR WHICH ELECTRONIC AND NUCLEAR CONTRIBUTIONS ARE NEGLIGIBLE

§1. *Introduction.* I have now everything I need for calculating the thermodynamic functions of an ideal gas of atoms and for comparing the results of such calculations to those obtained by measurements.

To simplify the theory, I have assumed that the molecules do not interact with each other. I have proven that if this is the case, then the equation of state is the ideal gas law. This means that the theory ought to agree with the experiment at those temperatures and pressures for which the p vs T data are reproduced well by the equation $pv = RT$. For noble gases, we are in this regime when $p \leq 1$ atm and $T \geq 298$ K.

For noble gases and hydrogen atoms, the excited electronic energies are much larger than $k_B T$, and the electronic ground state is not degenerate. For these reasons, the electronic degrees of freedom *do not contribute* to the thermodynamic functions of these gases. Thus, I only need to calculate the translational contributions.

If the theory is correct, it must allow me to calculate accurately all thermodynamic functions of He, Ar, Kr, Ne, Xe, H, etc. I will test if this is true for the case of argon.

The theory provides more than just a numerical recipe. It permits us to draw some qualitative conclusions that are also subject to verification. I will make a point of highlighting them.

To test the theory, it is sufficient to show that the Helmholtz free energy A calculated from thermodynamic data agrees with the one obtained from statistical mechanics. All other thermodynamic quantities are obtained by taking derivatives of A; if A agrees with experiments, all other functions will do so too. In spite of this, I calculate the entropy, the enthalpy, and the chemical potential of 1 mole of argon, to give you some practice in such calculations. They all agree very well with experiment.

§2. *Qualitative Observations.* Examine the equations obtained in Chapter 6, to make some simple, qualitative predictions. I derived the ideal gas law from a purely atomic theory. This alone is remarkable. I have learned that a gas is ideal when the interaction between molecules can be neglected.

I have shown that the energy of 1 mole of gas is $u = u_t = \frac{3}{2}k_B T$ (see Eq. 6.23). This says that u does not depend on volume or pressure. Joule performed experiments that show this to be true.

A more striking prediction can be made regarding C_v and C_p (see Eq. 6.46 and the text below it): $C_v = \frac{3}{2}R$ and $C_p = \frac{5}{2}R$. Theory predicts that all noble gases and a gas of hydrogen atoms have the same heat capacities, at all temperatures and pressures, as long as the gas is ideal. In the NIST database (webbook.nist.gov/chemistry), the C_p values of He, Ar, Ne, Kr, Xe, and H are 4.97 cal/mol K for any temperature between 298 K and 6000 K. Since $R = 1.98722$ cal/mol K, the theory predicts that $C_p = \frac{5}{2} \times 1.98722$ cal/mol K $= 4.96805$ cal/mol K. This is in remarkable agreement with the data given in the tables.

I must warn you that it is likely that the "data" in the NIST tables were calculated with the formulae given here. Therefore we test that we and the people who created the tables are able to implement the equations correctly (a cynic might say that we managed to make the same mistakes). However, extensive comparison with experiments have been performed between 1920–1940 and the agreement is excellent.

In his article on argon Din[a] gives C_p and C_v values up to 600 K, which also agree fairly well with this calculation. For example, between 300 K and 600 K, at 1 atm, the value for C_p is 20.8 J/mol K, or 4.97 cal/mol K.

The theory also predicts that if an atom has low-lying excited states, then C_p depends on temperature and varies from atom to atom. You will see in the next chapter that this prediction is also correct.

§3. *Quantitative Tests: Entropy.* The entropy per mole can be calculated from the first three terms of Eq. 6.39 (in which I take $N = N_A$ and $N_A k_B = R$):

$$s(T, V) = s_t = R \ln \frac{V}{N_A \Lambda^3} + \frac{5}{2} R + R \ln g_0^n \tag{7.1}$$

This is the entropy of 1 mole of gas at given V and T. g_0^n is the degeneracy of the nuclear state.

The "experimental" data are given at $p = 1$ atm. Therefore I use the ideal gas law $V = RT/p$ for 1 mole of gas in Eq. 7.1, to obtain

$$\begin{aligned} s(T, p) &= R \ln \frac{k_B T}{p \Lambda^3} + \frac{5}{2} R + R \ln g_0^n \\ &= R \ln \frac{RT}{p N_A \Lambda^3} + \frac{5}{2} R + R \ln g_0^n \end{aligned} \tag{7.2}$$

Note that we calculate the entropy of 1 mole of gas at given pressure and temperature.

The de Broglie thermal wavelength

$$\Lambda(T) = \sqrt{\frac{h^2}{2\pi m k_B T}}$$

has units of length. If I use $h = 6.6262 \times 10^{-27}$ erg s, $k_B = 1.3806 \times 10^{-16}$ erg/K, and m in grams (1 amu $=1.6605 \times 10^{-24}$ g), I obtain Λ in centimeters.

[a] *Thermodynamic Functions of Gases, Vol. 2*, F. Din, editor, Butterworths Scientific Publications, London, 1957, pp. 146–201.

In Workbook SM7.2, I calculate that

$$\Lambda(T) = 1.7459 \times 10^{-7} \sqrt{\frac{1}{mT}} \quad \text{cm}$$

where m is in amu (e.g. $m = 2$ for He) and T is in kelvin. The order of magnitude of Λ is around 1 Å or 10^{-8} cm.

As usual, the units can give one a headache. The argument of the logarithm must be dimensionless. Since Λ^3 has units of volume, if I use CGS units, then Λ^3 is in cm^3. I must have RT/pN_A in the same units. For this, I use $R = 82.0578$ cm^3 atm/mol K, the pressure in atm, and Avogadro's number $N_A = 6.0222 \times 10^{23}$ mol^{-1}. Finally, the units of s are determined by the units used for R in front of the logarithm, and for this I use $R = 1.98722 \times 10^{-3}$ cal/mol K, because the data for entropy are in these units.

With these values, Eq. 7.2 leads to (see Workbook SM7.2)

$$s(T, V) = -2.47479 + 2.98 \ln m + 4.968 \ln T - 1.987 \ln p$$
$$+ 1.987 \times 10^{-3} g_0^n \quad \text{cal/mol K} \tag{7.3}$$

In this expression, T is in kelvin, p is in atm, and m is dimensionless (e.g. $m = 39.95$ for Ar).

The entropy is weakly dependent (logarithmically) on T, p, and m. The order of magnitude when $T \gg 300$ K is controlled by $4.968 \ln T$ cal/mol K.

I compared the results given by this expression to the measured values for argon in Table 7.1 (where I assumed that $g_0^n = 1$). The agreement with "experiment" is impressive.

The data given by Din (see p. 95) for entropy are at many pressures but only at temperatures up to 600 K. At 1 atm and 300 K, $s_{\text{Din}} = 154.6$ J/mol = 36.9 cal/mol; at 1 atm and 600 K, $s_{\text{Din}} = 169$ J/mol = 40.4 cal/mol. These experimental values are even closer to our calculated values than the NIST data.

§4. *The Change in Enthalpy.* The enthalpy per mole can be calculated from the definition (see your thermodynamics textbook)

$$h = u + pv \tag{7.4}$$

Table 7.1 A comparison of the calculated entropies to the experimental values for argon at $p = 1$ atm. The experimental data are taken from the NIST site (webbook.nist.gov/chemistry).

Temperature, T (K)	s_{calc} (cal/mol K)	s_{exp} (cal/mol K)	Percentage error
298.15	36.8234	37.01	−0.507
300.00	36.8541	37.04	−0.504
600.00	40.2977	40.48	−0.452
900.00	42.3120	42.50	−0.444
1200.00	43.7413	43.93	−0.431
1500.00	44.8498	45.04	−0.424
1800.00	45.7556	45.94	−0.403
2100.00	46.5215	46.71	−0.405
2400.00	47.1849	47.37	−0.392
2700.00	47.7700	47.96	−0.398
3000.00	48.2934	48.48	−0.386

Here v is the molar volume, and the ideal gas law gives $pv = RT$. For u, I use (see Chapter 6, §5)

$$u = u_t = \tfrac{3}{2}RT \tag{7.5}$$

Combining Eqs 7.4 and 7.5 gives

$$h_t = \tfrac{3}{2}RT + RT = \tfrac{5}{2}RT \tag{7.6}$$

The experimental data are given in kcal/mol; therefore I use $R = 1.98722 \times 10^{-3}$ cal/mol K.

The theory predicts that all ideal monoatomic gases, for which the electronic contribution can be neglected, have the same enthalpy. I checked in the NIST tables the values of $\Delta h = h(T) - h(298.15)$ for Ar and Xe. The two compounds have identical Δh at all temperatures. In Workbook SM7.7, I calculated $h(T) - h(298.15)$ for argon. The results are shown in Table 7.2, together with the measured values. The agreement is excellent. The discrepancy at 300 K is probably due to experimental error: $h(300)$ and $h(298.15)$ are very close and to get the difference accurately, the measured values must be accurate to at least three decimal places.

Table 7.2 A comparison of $h(T) - h(298.15)$ calculated from Eq. 7.6 to the experimental values given by NIST (webbook.nist.gov/chemistry).

Temperature, T (K)	$h(T) - h(298.15)$ for Ar (kcal/mol)		Percentage error
	Calculated	Experimental	
298.15	0	0	0
300.00	0.000919	0.01	−8.80
600.00	1.499610	1.50	-2.33×10^{-2}
900.00	2.990020	2.99	6.99×10^{-4}
1200.00	4.480440	4.48	9.73×10^{-3}
1500.00	5.970850	5.97	1.43×10^{-2}
1800.00	7.461270	7.46	1.70×10^{-2}
2100.00	8.951680	8.95	1.88×10^{-2}
2400.00	10.442100	10.44	2.01×10^{-2}
2700.00	11.932500	11.93	2.10×10^{-2}
3000.00	13.422900	13.42	2.18×10^{-2}

§5. *Chemical Potential.* This is the most important thermodynamic function. It can be calculated from its definition (see your thermodynamics textbook)

$$\mu_t = h_t - Ts_t \tag{7.7}$$

Other ways of calculating μ_t are possible, but this is the easiest route since I have already derived equations for $s_t(T,p)$ and $u_t(T)$ (Eqs 7.3 and 7.6). Using them in Eq. 7.7 gives (see Workbook SM7.8)

$$\mu(T,p) = T\left[7.44284 - 2.981 \ln m + 1.987 \ln p - 4.968 \ln T\right] \text{ cal/mol} \tag{7.8}$$

I have assumed that $g_0^n = 0$. In this equation, T is in kelvin, m is in amu, and p is in atm. A comparison of the calculated values and the experimental values is made in Table 7.3. The calculations were performed in Workbook SM7.9.

The NIST tables give the quantity $-(G(T) - H(298.15))/T \equiv X(T)$, where $G(T)$ is the Gibbs free energy per mole (which is the chemical potential) and $H(298.15)$ is

Table 7.3 A comparison, for argon, of the calculated values of $\mu(T) - \mu(298)$ in cal/mol (from Eq. 7.8) with the experimental values derived from the NIST website.

Temperature, T (K)	$\mu(T, 1\text{ atm}) - \mu(298.15, 1\text{ atm})$ (cal/mol)		Percentage error
	Calculated	Experimental	
298.15	0	0	0
300.00	−68.1517	−68.4685	−0.46
600.00	−11,700.1	−11,753.5	−0.45
900.00	−24,111.9	−24,227.5	−0.48
1200.00	−37,030.2	−37,193.5	−0.44
1500.00	−50,325	−50,540.5	−0.43
1800.00	−63,920	−64,205.5	−0.46
2100.00	−77,764.5	−78,089.5	−0.42
2400.00	−91,822.7	−92,213.5	−0.42
2700.00	−106,068	−106,523	−0.43
3000.00	−120,479	−120,965	−0.40

the enthalpy per mole at $T = 298.15$ K. The pressure is 1 atm. In our notation,

$$X(T, 1\text{ atm}) = \frac{\mu(T, 1\text{ atm}) - h(298.15\text{ K}, 1\text{ atm})}{T}$$

From this, I obtain

$$\mu(T, 1\text{ atm}) - \mu(298.15\text{ K}, 1\text{ atm}) = -T\,X(T, 1\text{ atm}) + 298.15X(298.15\text{ K}, 1\text{ atm})$$

The calculated and measured values of $\mu(T, 1\text{ atm}) - \mu(298.15, 1\text{ atm})$ are shown in Table 7.3. The agreement between them is excellent.

Exercise 7.1

The NIST website provides the following formula for Xe

$$C_p = A + Bt + Ct^2 + Dt^3 + \frac{E}{t^2}\ \text{cal/mol K}$$

with $t = T/1000$ for T in kelvin, $A = 4.967974$, $B = 1.780431 \times 10^{-7}$, $C = -4.898184 \times 10^{-8}$, $D = 2.549379 \times 10^{-9}$, $E = 5.975765 \times 10^{-9}$.

(a) Use thermodynamics to calculate $a(T, 1\text{ atm}) - a(298.15, 1\text{ atm})$, $u(T, 1\text{ atm}) - u(298.15, 1\text{ atm})$, $h(T, 1\text{ atm}) - h(298.15, 1\text{ atm})$, and $\mu(T, 1\text{ atm}) - \mu(298.15, 1\text{ atm})$.

(b) Compare the results to those calculated from statistical mechanics.

§6. *Summary.* The theory is remarkably successful. The calculations start with a very meager input: the definition of the partition function $Q = \sum_n \exp[-E_n/k_B T]$, the connection to thermodynamics $A = k_B T \ln Q$, and the ad hoc division by $N!$ (a step that could be derived, had we been more sophisticated). Quantum mechanics supplies the formulae for E_n and the rest is simple algebra and calculus. The only assumption is that the density of gas is so low and the temperature is so high that the interactions between molecules can be neglected. Theory tells us that this neglect is legitimate whenever the p vs T data are well-fitted by the ideal gas law $pv = RT$.

The excitation energy of the electrons in argon is very large, and the ground electronic states are not degenerate. Because of this, only translational motion contributes to thermodynamic quantities. We find that the theory agrees very well with experiment. The discrepancies are probably due to experimental error. The test of the theory for atoms whose electronic excitation energies are not much larger than $k_B T$ is made in the next chapter.

Exercise 7.2

Go to the NIST website (webbook.nist.gov/chemistry) and get $C_p(T)$, $s(T, 1\text{ atm})$, $h(T, 1\text{ atm}) - h(298.15\text{ K}, 1\text{ atm})$, and $\mu(T, 1\text{ atm}) - \mu(298.15\text{ K}, 1\text{ atm})$ for xenon. Use statistical mechanics to calculate these quantities. Compare your results to the data, for a wide range of temperatures.

Exercise 7.3

You know the thermodynamic functions C_p, s, h, and μ for a gas of hydrogen atoms. Derive simple equations that express the thermodynamic functions of a gas of deuterium atoms in terms of the thermodynamic functions of a gas of hydrogen atoms.

Exercise 7.4

(a) Calculate the thermodynamic functions of a gas of oxygen atoms by assuming that the energy of the electronic excitations is much larger than $k_B T$ and that the electronic and nuclear ground states are not degenerate. Get data from the NIST website (webbook.nist.gov/chemistry). Do the calculations agree with the experiments?

(b) From the nature of the disagreement in (a), demonstrate that the electronic contributions to the thermodynamic functions for oxygen cannot be neglected.

8

A TEST OF THE THEORY FOR A GAS FOR WHICH ELECTRONIC AND NUCLEAR DEGREES OF FREEDOM MATTER

Introduction

§1. *"Something is Rotten in the State of Denmark."* In the previous chapter we have neglected the effect of the electronic degrees of freedom. The resulting theory gave accurate results for the thermodynamic functions of noble gases and a gas of atomic hydrogen. However, this simplified theory has trouble with other atoms. For example, it predicts that $C_p = 4.97$ cal/mol K for all atomic gases, at all temperatures, as long as the gas is ideal (see Chapter 7). Measurements on Ar agreed with this statement, but those for a gas of O atoms do not. The NIST website gives, for oxygen atoms, $C_p = 5.22$ cal/mol K at 298 K, 4.97 cal/mol K at 1600 K, and 5.33 cal/mol K at 6000 K. Obviously, C_p changes with temperature. We encounter similar difficulties when examining a gas of I or F atoms.

It is likely that these errors come from the neglect of the electronic and nuclear contributions to the thermodynamic functions. In this chapter we examine more closely the electronic and nuclear degrees of freedom and find that they cannot be always neglected. Moreover, when they are included, the theory is in excellent agreement with the measurements.

§2. *When do we Need to Include the Electronic Degrees of Freedom?* Since all thermodynamic quantities are calculated from the Helmholtz free energy a, we can start our analysis of the role of the electronic degrees of freedom by examining the electronic contribution a_{e}. This is given by (see Chapter 6, §3, Eq. 6.11)

$$a_{\mathrm{e}} = -RT \ln \left(q_{\mathrm{e}}\right) \tag{8.1}$$

The electronic partition function is (see Chapter 5, §12, Eq. 5.32)

$$q_{\mathrm{e}} = \exp\left[-\frac{\varepsilon_0^{\mathrm{e}}}{k_B T}\right]\left(g_{\mathrm{e}}^0 + \sum_{i=1}^{n} g_i^{\mathrm{e}} \exp\left[-\frac{\varepsilon_i^{\mathrm{e}} - \varepsilon_0^{\mathrm{e}}}{k_B T}\right]\right) \tag{8.2}$$

Here $\varepsilon_i^{\mathrm{e}}$ are the electronic energies of the atom and g_i^{e} is the degeneracy of the electronic state i. These quantities are given in Table 5.2, for several atoms.

If I insert Eq. 8.2 in Eq. 8.1, I obtain

$$a_{\mathrm{e}} = \frac{R\varepsilon_0^{\mathrm{e}}}{k_B} - RT \ln\left(g_0^{e} + \sum_{i=1}^{n} g_i^{\mathrm{e}} \exp\left[-\frac{\varepsilon_i^{\mathrm{e}} - \varepsilon_0^{\mathrm{e}}}{k_B T}\right]\right) \tag{8.3}$$

The first term in this equation is a constant (i.e. does not depend on T, V, or p) so we can discard it, since the free energy is defined up to an arbitrary constant. As a result, the electronic free energy is

$$a_{\mathrm{e}} = -RT \ln\left(g_0^{\mathrm{e}} + g_1^{\mathrm{e}} \exp\left[-\frac{T_1^{\mathrm{e}}}{T}\right] + g_2^{\mathrm{e}} \exp\left[-\frac{T_2^{\mathrm{e}}}{T}\right] + \cdots\right) \tag{8.4}$$

The quantity

$$T_i^{\mathrm{e}} \equiv \frac{\varepsilon_i^{\mathrm{e}} - \varepsilon_0^{\mathrm{e}}}{k_B} \tag{8.5}$$

has units of temperature and is called the *temperature of the electronic state i.*

We are now ready to take a closer look at the electronic free energy. We label the electronic energies so that $\varepsilon_0^{\mathrm{e}} \leq \varepsilon_1^{\mathrm{e}} \leq \varepsilon_2^{\mathrm{e}} \ldots$. This means that $T_1^{\mathrm{e}} \leq T_2^{\mathrm{e}} \leq \ldots$ and $\exp\left(-T_1^{\mathrm{e}}/T\right) \geq \exp\left(-T_2^{\mathrm{e}}/T\right) \ldots$. I also note that the magnitude of the degeneracies

g_i^e is between 1 and 5. As a consequence of these facts, we can neglect, in the sum appearing in Eq. 8.4, all exponentials for which

$$T_i^e \gg T \tag{8.6}$$

If this condition is satisfied by T_1^e then all exponentials in Eq. 8.4 can be ignored and the electronic free energy becomes

$$a_e = -RT \ln\left(g_0^e\right) \qquad \text{if } T_1^e \gg T \tag{8.7}$$

This is the equation used in Chapter 7, and it is correct for all gases for which $T_1^e \gg T$; if, in addition, $g_0^e = 1$ then the electronic free energy is equal to zero.

The electronic temperatures for a few atoms were calculated in Workbook SM8.1 and are given in Table 8.1.

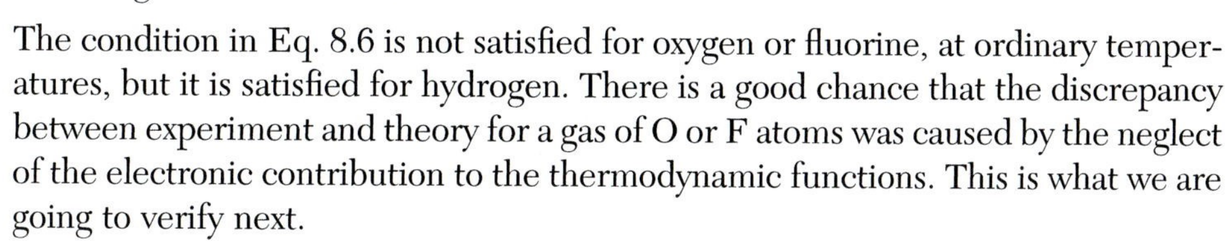

The condition in Eq. 8.6 is not satisfied for oxygen or fluorine, at ordinary temperatures, but it is satisfied for hydrogen. There is a good chance that the discrepancy between experiment and theory for a gas of O or F atoms was caused by the neglect of the electronic contribution to the thermodynamic functions. This is what we are going to verify next.

Exercise 8.1

Use the type of analysis performed above to determine whether the electronic degrees of freedom may affect some thermodynamic functions even at temperatures for which $T_1^e \gg T$.

Table 8.1 The electronic temperature and electronic degeneracy of several electronic states.

Atom	$\varepsilon_1-\varepsilon_0$ (eV)	T_1 (K)	$\varepsilon_2-\varepsilon_0$ (eV)	T_2 (K)	$\varepsilon_3-\varepsilon_0$ (eV)	T_3 (K)
O	0.02	232.8	0.03	348.12	1.97	22859.9
F	0.05	580.2	12.70	147,371		
H	10.20	118,361				

Exercise 8.2

How many exponentials would you keep in the sum calculating a_e for oxygen, fluorine, or hydrogen, at $T = 40$ K?

Exercise 8.3

(a) Show that the analysis performed for the electronic degrees of freedom is general (i.e. it is valid regardless of what the degrees of freedom are).

(b) If the analysis is general explain why we did not apply it to the translational degrees of freedom. Why don't we ever neglect the translational degrees of freedom?

Comparison with Experiment: the Method of Computation and the Results

§3. I want to calculate the heat capacity at constant pressure, the entropy, and the chemical potential of a gas of oxygen atoms and test whether including the contribution from the electronic degrees of freedom leads to results that agree with the data in the NIST tables.

As I explained in Chapter 1, once we know the Helmholtz free energy we can calculate all other thermodynamic functions by using Eqs 1.6–1.18. The equations we will use here are:

$$s = -\left(\frac{\partial a}{\partial T}\right)_{N,V} \tag{8.8}$$

for the entropy per mole, and

$$C_v = -\left(\frac{\partial u}{\partial T}\right)_{N,V} \tag{8.9}$$

together with

$$u = -T^2 \frac{\partial}{\partial T}\left(\frac{a}{T}\right)_{N,V} \tag{8.10}$$

for the heat capacity at constant volume (u is the energy of 1 mole of gas). The heat capacity at constant pressure is given by

$$C_p = C_v + R \tag{8.11}$$

Finally, the chemical potential μ is calculated from the definition

$$\mu = h - Ts = u + RT - Ts \tag{8.12}$$

To obtain the second equality, I used the definition of the enthalpy h, $h = u + pv$, and the ideal gas law.

The Helmholtz free energy per mole, needed in the equations given above, is (use Eq. 6.14 with $Nk_B \equiv R$ and replace $a_e = -RT \ln(q_e)$ with the expression given by Eq. 8.4):

$$\begin{aligned} a = & -RT \ln\left(\frac{Ve}{N_A \Lambda(T)^3}\right) \\ & - RT \ln\left(g_0^e + g_1^e \exp\left[-\frac{T_1^e}{T}\right] + g_2^e \exp\left[-\frac{T_2^e}{T}\right] + \cdots\right) \\ & - RT \ln\left(g_0^n\right) \end{aligned} \tag{8.13}$$

§4. *Implementation.* Eqs 8.8–8.13 provide a straightforward, but fairly tedious, procedure for calculating s, C_p, and μ. Luckily, the symbolic manipulation software (**Mathematica**, **Mathcad**) "knows" how to take derivatives, and I can instruct the computer to do all the hard work. I start by writing a program in which I define a function $a(T, V)$ and then ask the computer to perform all the derivatives required by the equations given above. This will generate expressions for all thermodynamic functions that I want to compute. Then, I can evaluate the magnitude of these expressions for $p = 1$ atm and several temperatures, and check whether the results agree with the measurements.

There is only one subtlety in this procedure: the derivatives in these equations are taken by keeping V and N constant (not p and N). Therefore, I must make sure that when I take the derivatives the free energy a is a function of V and N, and not of p and N.

If you use a calculator or a spread-sheet program to perform the calculations, you need explicit expressions for the thermodynamic functions. These have been derived in Chapter 6 and are summarized in Supplement 8.1.

As usual, I must pay attention to units. The free energy is equal to $RTf(T,V)$ where $f(T,V)$ is a dimensionless quantity. By taking $R = 1.98722$ cal/mol K into this expression, I obtain the free energy in units of cal/mol. All quantities calculated from this free energy will be in units of calories: s will have units of cal/mol K, u of cal/mol, and C_v and C_p of cal/mol K.

The quantity $Ve/N_A\Lambda(T)^3$, which appears in the formula for a, must be dimensionless, since it is the argument of a logarithm. I use CGS units, so that $h = 60626 \times 10^{-27}$ erg s and $k_B = 1.3806 \times 10^{-16}$ erg/K. The mass of the oxygen atom is $m = 16 \times 1.6605 \times 10^{-24}$ g. T is in kelvin. With these choices, Λ has units of centimeters and therefore V must be given in cm^3. Since we know the pressure and the temperature, the volume V of 1 mole of gas is calculated from the ideal gas law $V = RT/p$. If you use $R = 82$ cm^3 atm/mol K, you obtain V in cm^3.

The free energy a is calculated in Workbook SM8.2. The values of g_i^e and ε_i^e are taken from Table 5.2.

Exercise 8.4

Plot a_e of a gas of O atoms as a function of temperature (for $p = 1$ atm) and determine its value at the lowest temperature and the temperature where the electronic free energy starts to change. Use what you have learned in §2 to explain your results.

Exercise 8.5

Perform the same calculations as in the previous exercise, for a gas of F atoms. Explain where the difference between oxygen and fluorine comes from.

§5. *The Entropy of a Gas of Oxygen Atoms: Comparison to Experiment.* The calculation of the entropy is described in Workbook SM8.3, using Eq. 8.8.

The results are shown in Table 8.2. The agreement with the results given in the NIST database is excellent. The electronic contribution is about 10% of the total entropy.

§6. *The Heat Capacity.* In Workbook SM8.4, I used Eqs 8.9–8.11 to calculate the heat capacity at constant pressure C_p. The results are shown in Table 8.3 where they are compared with the data from the NIST website. The theory agrees very well with the experiment.

Table 8.2 A comparison of calculated entropy for a gas of oxygen atoms (see Workbook SM8.3) with the experimental values (from the NIST database). s_e is the electronic contribution.

Temperature, T(K)	s_e (cal/mol K)	s_{calc} (cal/mol K)	s_{exp} (cal/mol K)	Percentage error
198.15	4.20	38.3	38.49	−0.05
300.00	4.20	38.3	38.52	−0.49
600.00	4.32	41.9	42.08	−0.45
900.00	4.35	43.9	44.12	−0.43
1200.00	4.35	45.4	45.56	−0.42
1500.00	4.36	46.5	16.67	−0.40
1800.00	4.36	47.4	47.57	−0.38
2100.00	4.36	48.2	48.34	−0.38
2400.00	4.36	48.8	49.01	−0.38
2700.00	4.37	49.4	49.59	−0.37
3000.00	4.37	49.9	50.12	−0.37

Table 8.3 A comparison of the calculated heat capacity of a gas of oxygen atoms. The calculations were made in Workbook SM8.4. The experimental data are from the NIST database.

Temperature, T (K)	$(C_p)_{e,calc}$ (cal/mol K)	$(C_p)_{calc}$ (cal/mol K)	$(C_p)_{exp}$ (cal/mol K)	Percentage error
298.15	0.2820	5.25	5.23	0.3900
300.00	0.2800	5.25	5.23	0.3400
600.00	0.0863	5.05	5.06	−0.1100
900.00	0.0405	5.01	5.01	−0.0280
1200.00	0.0234	4.99	4.99	0.0280
1500.00	0.0152	4.98	4.98	0.0650
1800.00	0.0112	4.98	4.97	0.1900
2100.00	0.0104	4.98	4.97	0.1700
2400.00	0.0136	4.98	4.98	0.0340
2700.00	0.0220	4.99	4.99	0.0009
3000.00	0.0362	5.00	5.01	−0.1100

Table 8.4 A comparison of the calculated chemical potential with the measured values for a gas of oxygen atoms. The measurements are from the NIST database. The calculations were performed in Workbook SM8.5.

Temperature, T (K)	μ_{calc} (cal/mol)	μ_{exp} (cal/mol)	Percentage error
300	− 70.8831	−71.2065	−0.46
600	−12,168.7	−12,224.2	−0.46
900	−25,062.9	−25,181.2	−0.47
1200	−38,468.1	−38,636.2	−0.44
1500	−52,251.8	−52,484.2	−0.44
1800	−66,336.5	−66,66.2	−0.44
2100	−80,671.3	−81,008.2	−0.42
2400	−95,220.3	−95,612.2	−0.41
2700	−109,957	−110,402	−0.41
3000	−124,859	−125,354	−0.40

§7. *The Chemical Potential: Comparison to Experiment.* I used Eq. 8.12 to calculate the chemical potential. The results are shown in Table 8.4 together with the data from NIST data base. The agreement between theory and experiment is very good.

§8. *Summary.* These calculations show that the theory is capable of making quantitative predictions of all thermodynamic functions of an ideal gas of atoms. When the temperature of the gas is comparable to the electronic temperatures, the electronic contribution to a thermodynamic function can be as high as 10% of the total value.

Even at very low temperature, the electronic degrees of freedom may contribute to the entropy, if the electronic ground state is degenerate. Since the free energy and the chemical potential depend on entropy, they are also affected by ground-state degeneracy; the energy and the heat capacity are not.

Exercise 8.6

Calculate the thermodynamic functions for a gas of fluorine atoms. Compare your results to those given in the NIST database.

Exercise 8.7

Calculate the fraction of oxygen atoms that are in the first excited electronic state and plot it as a function of temperature.

Supplement 8.1 A Recipe for Calculating the Thermodynamic Functions when the Electronic and Nuclear Degrees are Included

If you do not use a symbolic manipulation program you will have to do the derivatives yourself. Fortunately, we have already done this in Chapter 6. I collect here those equations of Chapter 6 needed for calculating the heat capacity C_p, the entropy s, and the chemical potential μ.

§9. *The Heat Capacity.* In Chapter 6 we have seen that the heat capacity at constant volume is the sum of a translational and an electronic contribution:

$$C_v = (C_v)_\mathrm{t} + (C_v)_\mathrm{e} \tag{8.14}$$

The translational contribution is (see Eq. 6.53)

$$(C_v)_\mathrm{t} = \frac{3R}{2} \tag{8.15}$$

The electronic one is more complicated (see Eq. 6.55):

$$\frac{(C_v)_\mathrm{e}}{R} = \frac{1}{q_\mathrm{e}} \sum_{i\geq 1} g_i^\mathrm{e} \left(\frac{\varepsilon_i^\mathrm{e} - \varepsilon_0^\mathrm{e}}{k_B T}\right)^2 \exp\left[-\frac{\varepsilon_i^\mathrm{e} - \varepsilon_0^\mathrm{e}}{k_B T}\right]$$
$$- \frac{1}{q_\mathrm{e}^2} \left(\sum_{i\geq 1} g_i^\mathrm{e} \left(\frac{\varepsilon_i^\mathrm{e} - \varepsilon_0^\mathrm{e}}{k_B T}\right) \exp\left[-\frac{\varepsilon_i^\mathrm{e} - \varepsilon_0^\mathrm{e}}{k_B T}\right]\right)^2 \tag{8.16}$$

The electronic partition function q_e is given by Eq. 8.2. To calculate C_p, use $C_p = C_v + R$ which is a well-known equation from thermodynamics.

§10. *The Entropy.* Eq. 6.43 gives (use $N_A k_B = R$)

$$\frac{s}{R} = \frac{s_t + s_e + s_n + s_c}{R}$$
$$= \ln \frac{V}{N_A \Lambda(T)^3} + \frac{5}{2} + \ln g_0^n + \ln q_e(T) + \frac{1}{q_e} \sum_{i=1}^{\infty} \frac{g_i^e \varepsilon_i^e}{k_B T} \exp\left[-\frac{\varepsilon_i^e - \varepsilon_0^e}{k_B T}\right] \tag{8.17}$$

§11. *The Chemical Potential.* The chemical potential can be calculated from (use Eq. 6.65 and $pV = RT$)

$$\frac{\mu}{RT} = -\ln\left(\frac{RT}{p N_A \Lambda(T)^3}\right) - \ln q_e(T) - \ln g_0^n \tag{8.18}$$

The electronic contribution to the chemical potential is $-\ln q_e(T)$.

9

THE STATISTICAL MECHANICS OF A GAS OF DIATOMIC MOLECULES

Introduction

§1. *Outline.* In this chapter, I derive formulae giving the thermodynamic functions of an ideal gas of diatomic molecules. The presentation is similar to that used when we studied the thermodynamic functions of an ideal gas of atoms: we calculate first the partition function of the gas and use it to derive equations for the free energy A. Then we calculate all other thermodynamic functions from A, by using equations established by thermodynamics.

The energy of a diatomic molecule is the sum of translational, vibrational, rotational, electronic, and nuclear energies (see any book of quantum mechanics). Because of this, the partition function of the gas is a product of partition functions, each one corresponding to one type of motion. This, in turn, makes each thermodynamic quantity a sum of contributions from each type of motion.

Since we can calculate thermodynamic functions, we can use this theory for the same tasks as thermodynamics. The most important among these is the calculation of equilibrium constants and equilibrium concentrations in a chemical reaction (see Chapter 11). These calculations are particularly useful for combustion problems (e.g. rocket engines): the gases are ideal at high temperatures so the theory is accurate. The theory is particularly useful here because equilibrium measurements in a flame are very difficult to perform. If we were to use only phenomenological thermodynamics to study combustion, we would have to measure the heat capacities, the heats of formation, the equilibrium concentrations, the entropies, etc. at various temperatures, for species that are present in small amounts and only at very high temperatures. Such measurements are very difficult and sometimes impossible. By using statistical mechanics we can calculate these quantities from experimental information provided by spectroscopy: the vibrational frequency and rotational constant. These quantities are easier to measure than the thermodynamic ones. Moreover, nowadays we can obtain accurate values of vibrational frequencies and correct molecular structures, for small molecules, from quantum mechanical calculations, completely bypassing experiments.

The formulae obtained here permit us to study separately the contributions of translation, rotation, and vibration to various thermodynamic functions. By doing this we gain insight into the extent to which various kinds of molecular motions contribute to heat capacity, entropy, etc. The theory will also help us understand how the magnitude of these quantities varies from molecule to molecule and how they change with temperature. As an example, I examine some striking regularities of the heat capacities and explain how they are caused by differences in the vibrational energies.

In Chapter 10, I use the equations derived here to perform numerical calculations of entropy and heat capacity. This will give you a feeling for the order of magnitude of these quantities. A comparison with experiment will indicate that the approximations we make in this chapter fail at very high temperatures. This is not because the theory is incorrect, but because our treatment of vibrational and rotational motion is approximate, and these approximations deteriorate as the temperature is increased. The results obtained here are also used in Chapter 11 to calculate equilibrium constants and equilibrium compositions.

Finally, Supplement 9.1 explains the bizarre temperature dependence of the heat capacity of H_2 and D_2, at low temperatures, which disagree with the simple formulae derived in this chapter. The disagreement appears because we have ignored an important rule of quantum mechanics: when a molecule has identical nuclei, its wave function must change in a certain way when we exchange the nuclear positions. When this requirement is taken into account the theory leads to perfect agreement with the thermodynamic measurements.

The plan of the derivations presented in this chapter is very simple:

1. from quantum mechanics, get the energy eigenvalues for the vibrational and rotational energies of a diatomic molecule;
2. use these eigenvalues to calculate the partition function Q (Eq. 1.3);
3. from the partition function, calculate the Helmholtz free energy A (Eq. 1.5);
4. use Eqs 1.6–1.18 to calculate the thermodynamic functions.

We are also interested in the probability that a molecule is in a given state. This can be calculated from Eq. 1.2.

These calculations may be tedious (especially if you do not use a computer), but they are conceptually trivial.

§2. *The Partition Function.* We have seen in Chapter 4 that the partition function of a gas of non-interacting particles is (see Eqs 4.7–4.9)

$$Q = \frac{q^N}{N!} = \left(\frac{qe}{N}\right)^N \tag{9.1}$$

To obtain the second equality, I used Stirling's approximation, $N! \approx (N/e)^N$. The partition function q of one molecule is

$$q = \sum_{\alpha=0}^{\infty} \exp\left[-\beta \varepsilon_\alpha\right] \tag{9.2}$$

where ε_α is the quantum energy of the molecule alone in the box confining the gas and $\beta = 1/k_B T$. The sum is over all quantum states α of the molecule, not over all the energy values. I make this distinction because some quantum states can be degenerate: they have the same energy but differ through the values of other quantities.

You learned in quantum mechanics that the energy ε_α of a diatomic molecule is approximately given by

$$\varepsilon_\alpha = \varepsilon_\mathrm{t} + \varepsilon_\mathrm{v} + \varepsilon_\mathrm{r} + \varepsilon_\mathrm{n} + \varepsilon_\mathrm{e} \tag{9.3}$$

That is, the total energy ε_α of a diatomic molecule, alone in the box in which the gas is contained, is a sum of its translational energy ε_t, its vibrational energy ε_v, its rotational energy ε_r, the energy ε_n of the nuclei, and the energy ε_e of the electrons. This is an approximation: Eq. 9.3 ignores the coupling between the rotational and the vibrational energies. Had that coupling been included, we could not have separated the energy into a vibrational and a rotational part.

When I use Eq. 9.3 in the partition function q, given by Eq. 9.2, I obtain

$$\begin{aligned} q &= \sum_{\alpha} \exp[-\beta\varepsilon_{\alpha}] \\ &= \sum_{t,v,r,n,e} \exp[-\beta(\varepsilon_t + \varepsilon_v + \varepsilon_r + \varepsilon_n + \varepsilon_e)] \\ &= \sum_{t} \exp[-\beta\varepsilon_t] \sum_{v} \exp[-\beta\varepsilon_v] \sum_{r} \exp[-\beta\varepsilon_r] \\ &\quad \times \sum_{n} \exp[-\beta\varepsilon_n] \sum_{e} \exp[-\beta\varepsilon_e] \end{aligned} \tag{9.4}$$

I can write this as

$$q = q_t q_v q_r q_n q_e \tag{9.5}$$

where

$$q_t = \sum_{t} \exp[-\beta\varepsilon_t] \tag{9.6}$$

is the translational partition function,

$$q_v = \sum_{v} \exp[-\beta\varepsilon_v] \tag{9.7}$$

is the vibrational partition function,

$$q_r = \sum_{r} \exp[-\beta\varepsilon_r] \tag{9.8}$$

is the rotational partition function, and q_n and q_e are the nuclear and the electronic partition functions, defined similarly. The sums in these quantities are over all states of one molecule. All information about the quantities appearing in these sums is provided by quantum mechanics.

§3. *The Role of Electronic Excitations.* The inclusion of the electronic degrees of freedom for diatomic molecules is similar to that for atoms. There is, however, a complication. The length and the strength of the bond of an electronically excited molecule are different from those of the molecule in the ground electronic state. Because of this, the excited molecule has different vibrational and rotational energies than the ground-state molecule. This must be taken into account when the contribution from the excited electronic states is included.

As in the case of atoms (see Chapters 6 and 8), the effects of electronic excitation are important only at temperatures for which the magnitude of $k_B T$ is comparable to the excitation energy of the electrons (i.e. the energy of the excited state minus the energy of the ground state). For most molecules the electronic excitation energy is substantially larger than $k_B T$ and we can neglect the electronically excited state when we calculate the partition function. This means that, if the ground state is not degenerate, we can take

$$q_{\mathrm{e}} = 1 \tag{9.9}$$

Such a partition function has no effect on thermodynamic properties, since $\ln q_{\mathrm{e}} = 0$.

The oxygen molecule, which has a triplet ground state, is an exception to this rule. For it we must take

$$q_{\mathrm{e}} = 3 \quad \text{for } \mathrm{O}_2 \tag{9.10}$$

§4. *The Effect of Nuclear Degrees of Freedom.* The excitation of a nucleus requires a very high energy, and the nuclear degrees of freedom matter only when the ground state of the nucleus is *degenerate.* Such a degeneracy can appear because the nucleus can be in a variety of spin states. Their energies are so small and so close to each other that at ordinary temperatures we can consider them degenerate and having zero energy. If the spin of nucleus i is $I_i, i = 1, 2$, then the degeneracy is $(2I_1 + 1)(2I_2 + 1)$ and the nuclear partition function is

$$q_{\mathrm{n}} = (2I_1 + 1)(2I_2 + 1) \tag{9.11}$$

As you will see later, a temperature-independent partition function makes no contribution to energy and heat capacity, but does contribute to entropy, Helmholtz free energy, and Gibbs free energy. When we calculate the latter quantities we must take into account nuclear degeneracy.

Exercise 9.1

Prove the statements made in the paragraph above.

At low temperature, in the case of diatomics with identical nuclei, the nuclear wave function becomes "entangled" with the rotational one and the system acquires special properties. These effects become important when we study the thermodynamic functions of H_2 and D_2 at low temperature, and are discussed in Supplement 9.1.

§5. *The Thermodynamic Functions are Sums of Vibrational, Rotational, and Translational Contributions.* We have obtained here an interesting result: the partition function q of a molecule is a product of partition functions for vibration, rotation, translation, nuclear and electronic degrees of freedom. This happens because the total energy of the molecule is the sum of the vibrational, rotational, translational, electronic, and nuclear energies.

The decomposition of q into a product of partition functions has a very important consequence: since the Helmholtz free energy A is given by $A = -k_B T \ln Q$, when the partition function is a product of terms corresponding to different motions, A will be a sum of contributions from those motions.

To calculate the thermodynamic functions of a gas, I need to know the partition function Q. This is obtained by combining Eqs 9.5 and 9.1:

$$Q = \left(\frac{e\, q_{\mathrm{t}}\, q_{\mathrm{v}}\, q_{\mathrm{r}}\, q_{\mathrm{n}}\, q_{\mathrm{e}}}{N}\right)^N \tag{9.12}$$

The free energy of N molecules of gas is (see Eq. 9.1)

$$A = -k_B T \ln Q = -k_B T N \ln\left(\frac{e\, q_{\mathrm{t}}\, q_{\mathrm{v}}\, q_{\mathrm{r}}\, q_{\mathrm{n}}\, q_{\mathrm{e}}}{N}\right)$$

Hence

$$A = A_{\mathrm{t}} + A_{\mathrm{v}} + A_{\mathrm{r}} + A_{\mathrm{e}} + A_{\mathrm{n}} \tag{9.13}$$

The Helmholtz free energy A is a sum of contributions from various degrees of freedom.

The translational motion contributes

$$A_{\mathrm{t}} = -k_B T N \ln\left(\frac{e\, q_{\mathrm{t}}}{N}\right) \tag{9.14}$$

To reduce the number of equations, I have incorporated the collective contribution $-k_B Tln(e/N)$ into the translational free energy. Because of this, the translational contribution to free energy, entropy and chemical potential will be different from what they were in Chapter 5, where the collective contribution was treated separately.

The vibrational motion contributes

$$A_v = -k_B TN \ln(q_v) \tag{9.15}$$

The rotation adds

$$A_r = -k_B TN \ln(q_r) \tag{9.16}$$

The electronic contribution is

$$A_e = -k_B TN \ln(q_e) \tag{9.17}$$

Finally, the degeneracy of the nuclear ground state adds

$$A_n = -k_B TN \ln(q_n) \tag{9.18}$$

Because all other thermodynamic quantities can be obtained from A by taking derivatives, it follows that those quantities also consist of additive contributions from translational, rotational, vibrational, electronic, and nuclear degrees of freedom. For example, the entropy of the gas is the sum of a translational, vibrational, rotational, electronic, and nuclear entropies.

I remind you that Eq. 9.13 is obtained by assuming that the rotational and vibrational motions are decoupled and that electronic excitations are negligible. This approximation does not work well at high temperatures, for molecules that have soft bonds (their interatomic distance is easy to stretch).

Translational Contributions to Thermodynamic Quantities

§6. *The Translational Partition Function* q_t. When we studied a gas of N atoms we found that the translational partition function depends only on the mass of the particle. This means that Eq. 5.29 derived there can be used for a gas of diatomic molecules, as long as the mass is the total mass of the molecule $m = m_1 + m_2$. Thus we have (see Eq. 5.29)

$$q_{\mathrm{t}} = \frac{V}{\Lambda^3} \tag{9.19}$$

with

$$\Lambda = \sqrt{\frac{h^2}{2\pi m k_B T}} \tag{9.20}$$

Here Λ is the de Broglie thermal wavelength.

The translational contribution to the Helmholtz free energy

$$a_{\mathrm{t}} = -NRT \ln \left(\frac{Ve}{N_A \Lambda^3} \right) \tag{9.21}$$

where $R = k_B N_A$, with N_A the Avogadro number. Remember that this is the sum of the translational ($-k_B N_A \ln (V/\Lambda(T))$) and the collective $(-k_B T N_A \ln (e/N))$ contributions to the free energy. The translational energy per mole is (Eq. 6.23)

$$u_{\mathrm{t}} = -\tfrac{3}{2} RT \tag{9.22}$$

The translational entropy per mole is

$$s_{\mathrm{t}} = R \left[\tfrac{5}{2} + \ln \left(\frac{V}{N_A \Lambda^3} \right) \right] \tag{9.23}$$

This is the sum of Eqs 6.39 and 6.43. The translational chemical potential is (Eq. 6.67)

$$\mu_{\mathrm{t}} = -RT \ln \left(\frac{V}{N_A \Lambda^3} \right) \tag{9.24}$$

The translational heat capacity at constant volume is (Eq. 6.53)

$$(C_v)_{\mathrm{t}} = \tfrac{3}{2} R \tag{9.25}$$

And the translational heat capacity at constant pressure is (use $C_p = C_v + R$)

$$(C_p)_{\mathrm{t}} = \tfrac{5}{2} R \tag{9.26}$$

The Vibrational Contribution to Thermodynamic Quantities and the Population of the Vibrational States

§7. *No Measurable Quantity Changes if I Add a Constant to all Energies.* To calculate the vibrational partition function, in a form that is convenient for later applications, I will add a constant to the vibrational energies given by quantum mechanics. It is easy to show that adding a constant to all the vibrational energies does not affect any of the measurable predictions of the theory. To see this, I compare the results obtained by using the partition function

$$q_C = \sum_{i=1}^{\infty} \exp\left[-\frac{C + \varepsilon_i}{k_B T}\right]$$

to those obtained by using

$$q = \sum_{i=1}^{\infty} \exp\left[-\frac{\varepsilon_i}{k_B T}\right]$$

The only difference between the two is that I added the constant C to each energy in q_C. It is easy to see that $q_C = \exp[-C/k_B T]\, q$.

The Helmholtz free energy corresponding to the energies $\varepsilon_i + C$ is $A_C = -k_B T \ln q_C$; that corresponding to the energies ε_i is $A = -k_B T \ln q$. Using the expressions for q and q_C in these equations leads to $A_C = C + A$. Increasing each quantum energy ε_i by C increases the Helmholtz free energy by C.

Eqs 1.6–1.18 connect all thermodynamic quantities to the Helmholtz free energy. Thus, from A_C I can calculate the quantities corresponding to the energies $\varepsilon_i + C$, and from A I can calculate those corresponding to the energies ε_i. For example, the entropy S corresponding to the energies ε_i is $S = -(\partial A/\partial T)_V$ and the entropy S_C corresponding to the energies $\varepsilon_i + C$ is $S_C = -(\partial A_C/\partial T)_V = -(\partial (A + C)/\partial T)_V = S$. We find that the entropy is not affected by adding a constant to all the energies.

Exercise 9.2

Express all thermodynamic quantities calculated from statistical mechanics with the energies $\varepsilon_i + C$, denoted by A_C, U_C, S_C, H_C, G_C, $(C_p)_C$, $(C_v)_C$, and μ_C, in terms of the quantities A, U, S, H, G , C_p, C_v, and μ calculated with the energies ε_i. You will find that adding a constant to the energies adds a constant to A, U, H, and G, and leaves S, C_p, and C_v unchanged.

From this exercise, you see that when I change the energies from ε_i to $\varepsilon_i + C$, I must add a constant to A, U, H, and G. This seems to contradict the statement that a change from ε_i to $\varepsilon_i + C$ does not lead to a change that can be detected by experiments. There is no contradiction. All measurable quantities involve either derivatives of A, U, H, and G or differences between them (for example, the difference between U at a temperature T and U at 298 K). Adding a constant C to any of these functions (for example, to U) does not change their derivatives or the difference between the values of the function at two temperatures. It is therefore true that adding a constant to all the molecular energies in the partition function changes no measurable quantity.

This is a general property that has nothing to do with vibrational energies. We study it here in detail because it is the first time that we make use of it.

---

§8. *The Choice of the Vibrational Energy.* Quantum mechanics tells me that if the harmonic approximation is accurate, and if the rotational and vibrational motions are decoupled, then the vibrational energy of a molecule is

$$\varepsilon_{\mathrm{V}} = \hbar\omega \left(v + \tfrac{1}{2}\right) \tag{9.27}$$

Instead of this, I will use the equation

$$\varepsilon_{\mathrm{V}} = -D + \hbar\omega \left(v + \tfrac{1}{2}\right) \tag{9.28}$$

where D is defined by Fig. 9.1.

As I have argued, adding a constant to each vibrational energy does not make a difference in any measurable quantity. If this is so, why use Eq. 9.28 instead of Eq. 9.27? To understand the reason, look at Fig. 9.1, which shows the potential energy of a diatomic molecule as a function of the distance between the atoms. When the distance x between the atoms is large, there is no interaction between the atoms forming the molecule and the diatomic bond is broken. This fact is captured by the potential energy function: when x is large, $V(x)$ is a constant and the force $F = -\partial V/\partial x$ between atoms is zero. Note that adding a constant to V does not change the force ($\partial(V + C)/\partial x = \partial V/\partial x$).

The dashed-line potential $V(x)$ is equal to D when x is large and the atoms stop interacting. The potential $V(x) - D$ (the solid line) is zero when the atoms stop interacting. This choice of potential is consistent with the one used when we calculated

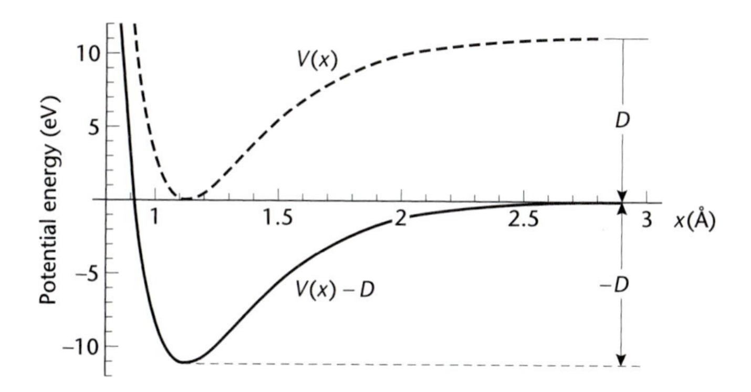

Figure 9.1 Potential energy of a diatomic molecule as a function of the distance x between the atoms. The dashed line is the potential energy $V(x)$ used in Quantum Mechanics. The solid line is the potential energy $V(x) - D$ used here.

the partition function of a gas of atoms. At that time we have assumed that the potential energy of the atoms in the box was zero.

It so happens that in quantum mechanics, it is customary to use the potential energy $V(x)$ shown as a dashed line in Fig. 9.1. In this case the energy scale is chosen so that *the lowest potential energy is zero.*

In the quantum theory of vibration, one makes the harmonic approximation to $V(x)$, which means

$$V(x) \simeq \frac{k}{2}(x - x_e)^2$$

where

$$k = \left(\frac{\partial^2 V}{\partial x^2}\right)_{x=x_e}$$

is the *force constant* and x_e is the *equilibrium position* of the diatomic. This is the position of the minimum of $V(x)$, i.e. the *bond length.* With this approximation, the eigenvalues of the energy of the harmonic oscillator are $\hbar\omega(v + \frac{1}{2})$, where

$$\omega = \sqrt{\frac{k}{\bar{m}}} \tag{9.29}$$

is the vibrational frequency, and

$$\bar{m} = \frac{m_1 m_2}{m_1 + m_2} \tag{9.30}$$

is the *reduced mass*.

Since the vibrational energy for $V(x)$ is $\hbar\omega(v+\frac{1}{2})$, the vibrational energy for $V(x)-D$ is $\varepsilon_v = \hbar\omega(v+\frac{1}{2}) - D$.

As long as I perform calculations for diatomic molecules alone, the addition or subtraction of D to the vibrational energy changes no measurable quantity of the system. I can use either Eq. 9.27 or Eq. 9.28. However, if I perform calculations in which I study a mixture of diatomic molecules AB and atoms A and B, in equilibrium, then all energies used in the calculations must be defined with the same energy scale. When I calculated the properties of a gas of atoms, the energy scale was chosen so that the lowest potential energy of the atoms was zero. Therefore I must use, for diatomics, an energy scale for which the potential energy of the atoms obtained by dissociating the diatomic molecule is zero. Only the potential $V(x) - D$ satisfies this condition. Therefore I must use $\hbar\omega(v+\frac{1}{2}) - D$ for the vibrational energy.

You will see, when we study chemical equilibrium, that the presence of D in the vibrational energy of the diatomic molecule is essential in obtaining the correct equilibrium constant. In fact, D dominates the physics of dissociation.

§9. *The Vibrational Partition Function.* The vibrational energy given by Eq. 9.28 can be written as

$$\varepsilon_v = -\left(D - \frac{\hbar\omega}{2}\right) + \hbar\omega v = -D_0 + \hbar\omega v \tag{9.31}$$

where $D_0 \equiv D - \hbar\omega/2$ is the dissociation energy of the molecule. The dissociation energy can be measured and it can be calculated by solving the Schrödinger equation for electrons.

I can now easily calculate the partition function:

$$q_{\rm v} = \sum_{v=0}^{\infty} \exp\left[-\frac{\varepsilon_{\rm v}}{k_B T}\right]$$

$$= \sum_{v=0}^{\infty} \exp\left[-\left(\frac{-D_0 + \hbar\omega v}{k_B T}\right)\right]$$

$$= \exp\left[\frac{D_0}{k_B T}\right] \sum_{v=0}^{\infty} \left(\exp\left[-\frac{\hbar\omega}{k_B T}\right]\right)^v \tag{9.32}$$

The last equality was obtained by using $\exp[a+b] = \exp[a]\exp[b]$ and $\exp[ab] = (\exp[a])^b$.

Using the notation

$$x = \exp\left[-\frac{\hbar\omega}{k_B T}\right]$$

in Eq. 9.32 gives me

$$q_{\rm v} = \exp\left[\frac{D_0}{k_B T}\right] \sum_{v=0}^{\infty} x^v \tag{9.33}$$

But I know (see, for example, Dwight[a]) that

$$\sum_{v=0}^{\infty} x^v = \frac{1}{1-x}$$

as long as $x^2 < 1$.

Using this fact in Eq. 9.33 gives

$$q_{\rm v} = \frac{\exp\left[D_0/k_B T\right]}{1-x} = \frac{\exp\left[D_0/k_B T\right]}{1-\exp\left[-\hbar\omega/k_B T\right]} \tag{9.34}$$

[a] H.B. Dwight, *Tables of Integrals and Other Mathematical Data*, MacMillan Publishing Co., Inc., New York, 1961, p. 3.

This is the vibrational partition function, within the harmonic approximation. The formula is valid only if $x^2 < 1$. But, since $\hbar\omega/k_BT$ is a positive number, $x^2 = \exp[-2\hbar\omega/k_BT]$ is always between 0 and 1, except at infinite temperatures; the formula is always valid.

Exercise 9.3

Test the equation $\sum_{n=0}^{\infty} x^n = \frac{1}{1-x}$ (valid for $x^2 < 1$) in several ways.

(a) Calculate the sum in the left-hand side numerically for several values of x and compare the result with that calculated from the right-hand side. Check what happens as x approaches 1 or is larger than 1.

(b) Perform the Taylor series expansion of $1/(1-x)$ and show that it is $\sum_{n=0}^{\infty} x^n$. If you are really good at math, prove that the series is convergent for $x^2 < 1$.

§10. *A Summary of the Approximations Made in Calculating* q_{v}. To obtain Eq. 9.34, I made the harmonic approximation and assumed that the rotational and the vibrational motion are decoupled. These two approximations are discussed in quantum mechanics. They break down when the molecule has high vibrational or rotational energy. Such energies contribute to the partition function only when the temperature of the gas is high.

Exercise 9.4

The vibrational energy eigenstates corrected for small deviations from the harmonic approximation are given by

$$\varepsilon_{\text{v}} = \hbar\omega(v + \tfrac{1}{2}) + \hbar\omega x_e(v + \tfrac{1}{2})^2$$

For I_2, the values determined experimentally are $\hbar\omega = 214.50\ \text{cm}^{-1}$ and $\hbar\omega x_e = 0.614$. The dissociation energy is 1.542 eV. Calculate, numerically, the partition function of I_2 with the energies ε_{v} given above and compare it to that given by the harmonic approximation. Go up to 1200 K. The data are from Huber and Herzberg.[b]

[b] K.P. Huber and G. Herzberg, *Molecular Spectra and Molecular Structure IV. Constants of Diatomic Molecules*, Van Nostrand, New York, 1979.

Table 9.1 The dissociation energy D_0 (see Eq. 9.31 and Fig. 9.1), the vibrational temperature $T_v = \hbar\omega/k_B$, and the rotational temperature $T_r = B/k_B$ for several diatomic molecules, for the more abundant isotopes.

Molecule	D_0 (KJ/mol)	T_r(K)	T_v(K)
H_2	432.073	87.5470	6338.20
N_2	941.400	2.8750	3392.01
O_2	491.888	2.0790	2273.64
Cl_2	239.216	0.3450	807.30
I_2	148.810	0.0537	308.65
HCl	427.772	15.2340	4301.38
HI	294.670	9.3690	3322.24
CO	1070.110	2.7770	3121.48
NO	627.700	2.4520	2738.87
Na_2	72.380	0.2210	229.00
K_2	66.110	0.0810	133.00

§11. *The Vibrational Temperature.* The relative motion of the atoms in a diatomic molecule is characterized by two energies: the dissociation energy D_0 and the vibrational quantum $\hbar\omega$ of the harmonic oscillator. These two energies do not appear isolated in the partition function. They are always divided by the energy $k_B T$. Since the Boltzmann constant k_B has units of energy divided by temperature (in kelvin), an energy divided by k_B has units of temperature. The quantity $T_v = \hbar\omega/k_B$ is called the *vibrational temperature*. Since ω is a property of the molecule, so is T_v; each diatomic molecule has its own vibrational temperature (see Table 9.1 for a few values). With this notation the vibrational partition function can be written as

$$q_v = \frac{\exp[D_0/k_B T]}{1 - \exp[-T_v/T]} \tag{9.35}$$

Exercise 9.5

(a) Prove that

$$q_v \approx \exp\left[\frac{D_0}{k_B T}\right]\frac{T_v}{T} \qquad \text{if } T \gg T_v \tag{9.36}$$

$$q_v \approx \exp\left[\frac{D_0}{k_B T}\right] \qquad \text{if } T \ll T_v \tag{9.37}$$

(b) Show that if $T \ll T_v$ and Eq. 9.37 is correct, then the vibrational motion q_v does not contribute to any measurable quantity.

Exercise 9.6

Denote by A_v, U_v, and S_v the vibrational contributions to the free energy, energy, and entropy, respectively. Calculate $A_v(T) - A_v(298)$, $U_v(T) - U_v(298)$, $S_v(T)$, and the contribution of vibration to the heat capacity C_v when $T_v \gg T$. Show that all these quantities are zero.

Exercise 9.7

Show that if $T \gg T_v$ then $A_v = -k_B TN \ln(k_B T/\hbar\omega)$, $U_v = \frac{1}{2}Nk_B T$, $(C_v)_v = k_B N_A$, and $S_v = k_B TN \ln(k_B T/\hbar\omega) + k_B N$.

Exercise 9.8

Show that the contribution to the pressure, due to vibrational and rotational motion, is zero. Explain why this is reasonable.

§12. *Vibrational Contribution to the Thermodynamic Functions.* I can now apply the equations given in Chapter 1 to calculate the vibrational contributions to various thermodynamic quantities. For a gas containing N diatomic molecules, I have (see Eq. 9.15) $A_v = -k_B TN \ln(q_v)$. Inserting in this the expression (9.35) for q_v gives

$$\begin{aligned} A_v &= -k_B TN \ln\left(\frac{\exp[D_0/k_B T]}{1 - \exp[-T_v/T]}\right) \\ &= -ND_0 + k_B NT \ln\left(1 - \exp\left[-\frac{T_v}{T}\right]\right) \end{aligned} \tag{9.38}$$

(I used $\ln e^x = x$ and $\ln(1/y) = -\ln y$.) If you want the free energy per mole, replace N with N_A (Avogadro's number) and use $k_B N_A = R$ (where R is the gas constant). The free energy per mole is therefore

$$a_v = -N_A D_0 + RT \ln\left(1 - \exp\left[-\frac{T_v}{T}\right]\right) \tag{9.39}$$

It is easy now to calculate the vibrational contribution to all other thermodynamic quantities, from A_v (or a_v), using Eqs 1.6–1.18. I will not give details, since the calculations are straightforward applications of calculus. I give only the starting equations and the results. The results are derived in Workbook SM9.1.

- The vibrational internal energy per mole (use Eqs 9.39 and 1.9) is:

$$u_v = -T^2 \frac{\partial}{\partial T}\left(\frac{a_v}{T}\right)_{N,V} = -N_A D_0 + \frac{RT_v}{\exp[T_v/T] - 1} \tag{9.40}$$

- The vibrational contribution to the heat capacity at constant volume is (use Eqs 9.40 and 1.11):

$$(C_v)_v = \left(\frac{\partial u_v}{\partial T}\right)_{N,V} = \frac{R(T_v/T)^2 \exp[T_v/T]}{(\exp[T_v/T] - 1)^2} \tag{9.41}$$

- The vibrational contribution to the entropy per mole is (use Eqs 9.39 and 1.6):

$$s_v = -\left(\frac{\partial a_v}{\partial T}\right)_{V,N} = \frac{RT_v/T}{\exp[T_v/T] - 1} - R\ln\left(1 - \exp\left[-\frac{T_v}{T}\right]\right) \tag{9.42}$$

- The vibrational contribution to the chemical potential, per molecule, is (use Eqs 9.39 and 1.16):

$$\mu_v = \left(\frac{\partial A_v}{\partial N}\right)_{T,V} = -D_0 + k_B T \ln\left(1 - \exp\left[-\frac{T_v}{T}\right]\right) \tag{9.43}$$

To obtain the chemical potential per mole, multiply by Avogadro's number N_A.

Exercise 9.9

Show that as T_v/T becomes very large, a_v approaches $-N_A D_0$, u_v approaches $-N_A D_0$, $(C_v)_v$ approaches 0, s_v approaches 0, and μ_v approaches $-D_0$. Use the information-theory approach discussed in Chapter 3 to explain why $s \to 0$ when $T_v/T \to \infty$.

Exercise 9.10

Calculate $a_v(T) - a_v(298)$, $u_v(T) - u_v(298)$, $s_v(T)$, $\mu_v(T) - \mu_v(298)$, and $(C_v)_v$ for K_2 and H_2 at pressure $p = 1$ atm and temperatures T varying from 400 K to 2000 K in increments of 200 K. Make a table with the results for K_2 and H_2.

§13. *The Probability of Being in State* v. The population of the excited vibrational states affects the absorption and emission spectra and sometimes the reactivity of the diatomic molecules. This is why we need to know what fraction of the total number of molecules in the gas is in state $v = 1$, or $v = 2$, etc. where v is the vibrational quantum number.

One of the fundamental equations presented in Chapter 1 (see Eq. 1.2) stated that the probability that a molecule is in state v is

$$P_v = \frac{\exp[-\varepsilon_v/k_B T]}{\sum_{v'} \exp[-\varepsilon_{v'}/k_B T]} = \frac{\exp[-\varepsilon_v/k_B T]}{q_v} \tag{9.44}$$

To obtain the second equality, I used Eq. 9.7, which defines q_v. From Eq. 9.35, $q_v = \exp[D_0/k_B T]/\left(1 - \exp[-T_v/T]\right)$. Eq. 9.31 gives $\varepsilon_v = -D_0 + \hbar\omega v$. Using these two results in Eq. 9.44 gives

$$P_v = \frac{\exp\left[\dfrac{D_0}{k_B T}\right]\exp\left[\dfrac{-\hbar\omega v}{k_B T}\right]}{\exp\left[\dfrac{D_0}{k_B T}\right]\Big/\left(1 - \exp\left[-\dfrac{T_v}{T}\right]\right)}$$

$$= \exp\left[-\frac{vT_v}{T}\right]\left(1 - \exp\left[-\frac{T_v}{T}\right]\right) \tag{9.45}$$

To obtain the last equality I used the definition $T_v = \hbar\omega/k_B$. Note that D_0 does not affect P_v.

There is a subtlety here. The statement above is true only if we assume that the molecule does not dissociate. If D_0 is small and the temperature is high, some of the diatomic molecules will be dissociated. In this case we need to be more careful how we define the probability that a diatomic that has not dissociated, and is in equilibrium with a gas of atoms, is in a state v.

If the temperature is such that $T_v/T \gg 1$, then $P_v \approx 0$ for $v = 1, 2, \ldots$, and $p_0 \approx 1$; practically all the molecules are in the ground state. As T increases toward T_v and exceeds it, the population of the excited states increases.

Exercise 9.11

For Br_2 and K_2, plot the probabilities P_0, P_1, and P_2 as functions of temperature.

Exercise 9.12

The vibrational energy of an anharmonic molecule is

$$E_v = \hbar\omega(v + \tfrac{1}{2}) + \hbar\omega x_e(v + \tfrac{1}{2})^2$$

For Pb_2, $\hbar\omega = 256.5\ \text{cm}^{-1}$, $\hbar\omega x_e = 2.96\ \text{cm}^{-1}$, and $D_0 = 0.7$ eV. (The Data are from Herzberg[c]) Calculate a_v, u_v, s_v, $(C_v)_v$, $(C_p)_v$, and μ_v with and without the correction $\hbar\omega x_e(v + \frac{1}{2})^2$. Is the correction important? Does it matter more when T is low or when T is high? Explain your reasoning. *Note*: To calculate the partition function with the anharmonic correction, you must evaluate numerically the sum $q_v = \sum_{v=0}^{\infty} \exp[-E_v/k_B T]$.

The Rotational Contribution to Thermodynamic Functions

§14. *The Rotational Energy.* When you studied quantum mechanics you learned that the energy levels of a rotating diatomic molecule are

$$\varepsilon_r(j) = j(j+1)B, \quad j = 0, 1, 2, \ldots \tag{9.46}$$

Here B is the rotational constant:

$$B = \frac{\hbar^2}{2\bar{m}r_e^2} \tag{9.47}$$

where r_e is the bond length of the molecule, $\bar{m}$ is the reduced mass (see Eq. 9.30), and $\hbar = 1.05457266 \times 10^{-34}$ J s is the "new" Planck constant (the "old" Planck constant, h, is equal to $2\pi\hbar$).

[c] G. Herzberg, *Molecular Spectra and Molecular Structure I. Spectra of Diatomic Molecules*, Van Nostrand Reinhold Co., New York, 1950.

You have also learned that the rotational states are degenerate: for a given value of j, there are $2j + 1$ states with the energy given by Eq. 9.46. These $2j + 1$ states correspond to different orientations of the rotating molecule.

It turns out that this formula for the rotational energy is valid only for *heteronuclear* diatomics (i.e. diatomics with different nuclei, such as HD, $^{16}O^{18}O$ or NO). The equations for *homonuclear* diatomics (the ones in which the nuclei are identical down to the last neutron, such as H_2 or $^{16}O^{16}O$) are different and will be discussed in Supplement 9.1.

You should read that material to see how some exotic requirements of quantum mechanics lead to a complete explanation of the very bizarre behavior of the thermodynamic properties of H_2, D_2, and HD at low temperature.

§15. *The Rotational Partition Function.* The rotational partition function is (see Eq. 9.8)

$$q_r = \sum_{j=0}^{\infty} (2j+1) \exp\left[-\frac{\varepsilon_r(j)}{k_B T}\right] = \sum_{j=0}^{\infty} (2j+1) \exp\left[-\frac{j(j+1)B}{k_B T}\right] \tag{9.48}$$

The factor $2j+1$ appears in Eq. 9.48 for the following reason: the sum in the partition function is over all states, not over all energies. For each value of j, there are $2j + 1$ states having the energy $j(j + 1)B$, and therefore each term $\exp\left[-j(j+1)B/k_B T\right]$ appears $2j + 1$ times in the sum.

The rotational constant B has units of energy and therefore $T_r = B/k_B$ has units of temperature; it is called *the rotational temperature of the molecule*. With this notation, q_r becomes

$$q_r = \sum_{j=0}^{\infty} (2j+1) \exp\left[-j(j+1)\frac{T_r}{T}\right] \tag{9.49}$$

The values of the rotational temperature T_r for several diatomic molecules are given in Table 9.1. As you can see, T_r is fairly low: the highest value, for H_2, is 87.547 K. Therefore, for most problems of interest to chemists, $T \gg T_r$. This allows us to turn the sum in Eq. 9.49 into an integral and evaluate it to obtain a formula for q_r. This can be done in two ways. One follows the argument presented when we derived the translational partition function for a gas of atoms (see Chapter 5, §6). The other uses the Euler–Maclaurin formula, which connects a sum to an integral.

The result obtained by using the first procedure is

$$q_r = \sum_{j=0}^{\infty}(2j+1)\exp\left[-j(j+1)\frac{T_r}{T}\right]$$
$$= \int_0^{\infty}(2x+1)\exp\left[-x(x+1)\frac{T_r}{T}\right]dx \qquad (9.50)$$

The integral is very easy to evaluate (see Workbook SM9.2). If you want to do it yourself, substitute $y = x(x+1)$ and $dy = (2x+1)dx$. The integral becomes

$$\int_0^{\infty} e^{-yT_r/T}dy = \frac{T}{T_r}$$

Therefore

$$q_r = \frac{T}{T_r} \quad \text{if } T_r \ll T \qquad (9.51)$$

The same result can be derived by using the Euler–Maclaurin formula, which is

$$\sum_{j=0}^{\infty} f(i) = \int_0^{\infty} f(x)dx + \frac{1}{2}\left[f(\infty) - f(0)\right]$$
$$+ (-1)\frac{B_1}{2!}\left[f^{(1)}(0) - f^{(1)}(\infty)\right]$$
$$+ (-1)^2\frac{B_2}{4!}\left[f^{(3)}(0) - f^{(3)}(\infty)\right]$$
$$+ (-1)^3\frac{B_3}{6!}\left[f^{(5)}(0) - f^{(5)}(\infty)\right] + \cdots$$

(See the bible of numerical calculations[d].) This equation gives a more accurate evaluation of the sum in Eq. 9.49. Using it (Workbook SM9.3) leads to

$$q_r = \frac{T}{T_r}\left[1 + \frac{1}{3}\frac{T_r}{T} + \frac{1}{15}\left(\frac{T}{T_r}\right)^2 + \frac{4}{315}\left(\frac{T}{T_r}\right)^3 + \cdots\right] \quad \text{if } T_r \ll T \qquad (9.52)$$

[d] W.H. Press, S.A. Teukolsky, W.I. Vetterling, and B.P. Flannery, *Numerical Recipes in C*, Cambridge University Press, 1992, p. 138.

If you compare Eq. 9.52 to Eq. 9.51, you see that the leading term in Eq. 9.52 is equal to the result given by Eq. 9.51. The remaining terms in Eq. 9.52 are corrections to Eq. 9.51, which can be neglected if

$$\frac{T}{T_r} \gg \frac{1}{3} \tag{9.53}$$

There is a third alternative in calculating q_r: use a computer and perform the sum in Eq. 9.49. Computers are now so powerful, that I don't bother anymore with Eqs 9.51 and 9.52; I just add up the first 300 or so terms in the sum in Eq. 9.49. To check whether I used enough terms I repeat the calculation with 350 terms, to test whether the result changes. If it does not, 300 terms gave me the correct result.

Exercise 9.13

Calculate the temperature T above which you can safely use the formula $q_r = T/T_r$, for I_2, H_2, and HCl (use Eq. 9.53).

Exercise 9.14

Calculate q_r for HCl at 10 K and at 1000 K by calculating the sum that defines q_r. Compare the results to the formulae in Eqs 9.51 and 9.52. *Note*. HCl is a solid at 10 K, so this is an exercise in "pedagogical physics"; we pretend that it is a gas, to illustrate some mathematical aspects of performing sums.

---

§16. *The Rotational Partition Function for Homonuclear Diatomics.* It has been observed that Eq. 9.49 works very well for heteronuclear diatomics (the two nuclei are different), but that it had difficulties with homonuclear molecules (the nuclei are identical). It was found empirically that to get the correct rotational contribution to the entropy of a gas of homonuclear diatomic molecules, one must divide the partition function by 2. This means that

$$q_r = \frac{T}{\sigma T_r} \quad \text{for } T \gg T_r \tag{9.54}$$

where $\sigma = 1$ for heteronuclear diatomics and $\sigma = 2$ for homonuclear diatomics. This formula is valid if $T \gg T_r$ and therefore covers most situations of interest to chemists.

The *symmetry factor* σ was introduced in the early days of statistical mechanics because without it the entropy of a gas of homonuclear diatomics differed from the measured one. The factor of 2 fixed the problem, just as the division by $N!$ fixed the entropy of the translational degrees of freedom. In neither case was there a theoretical justification for the division.

Ingenious people can find a plausible explanation for everything, and a factor of 2 is never much of a challenge. You can still find in some of the older books (which remain excellent reading) the following explanation. Imagine that you put an axis through the middle of a homonuclear molecule (perpendicular to the bond) and rotate the molecule by 180°. A homonuclear molecule after the rotation is indistinguishable from the molecule before the rotation and a heteronuclear one is not. Therefore, the argument went, by not taking into account this fact we have included in the partition function twice the number of states that we should. To fix this error we should divide by 2.

The argument took hold, except with those inclined to generalizations. What should we do for benzene? If I rotate the molecule around an axis perpendicular to the molecule and going through its center by 360/6 degrees, I bring the molecule to a state identical to that prior to the rotation. Should I divide the partition function by 6? But how about the rotations around the axes going through the carbon atoms on the opposite sides or the ring. Should those rotations contribute a factor of 2? All three of them?

As the deeper minds pondered such questions, experiments demolished the division by 2. It turned out that if $T \lesssim T_r$, Eq. 9.54 fails miserably. This can be seen by calculating the thermodynamic functions for H_2 and comparing them to experiment. The disagreement is qualitative: the curve showing how C_p changes with temperature has the wrong shape. This disagreement is even more striking if we note that the calculations for HD (where D is deuterium), using Eq. 9.54 with $\sigma = 1$ (HD is heteronuclear) agree very well with experiment. At first sight, this makes no sense. HD differs from H_2 by one neutron only! Why in the world would adding a neutron to a nucleus make such a dramatic difference in the magnitude and the temperature dependence of, say, C_p? To make things more mysterious, the calculations for D_2 also disagree with experiment and the results are very different from those for HD or H_2. Adding another neutron, to go from HD to D_2, again causes a radical change in the thermodynamic functions of a compound. If we wanted to keep σ in Eq. 9.54 and bring the equation into agreement with the facts, we would have to give σ a very complicated temperature dependence. That makes no sense.

Why would such differences between the properties of very similar molecules exist? Adding a neutron changes the mass. This changes the vibrational frequency and the rotational constant, but taking that into account does not bring the calculation into agreement with experiment. We need to take a different and much deeper route.

The disagreement with experiment appears only when the nuclei are identical. When you studied quantum mechanics, you learned that the wave function of identical particles is subject to special requirements. If the spin of the identical particles is half-integer (i.e. $\frac{1}{2}$, $\frac{3}{2}$, etc.), the wave function must be anti-symmetric ($\Psi(\mathbf{r}_1, \mathbf{r}_2) = -\Psi(\mathbf{r}_2, \mathbf{r}_1)$). If their spin is an integer (i.e. 1, 2, etc.), then the wave function must be symmetric ($\Psi(\mathbf{r}_1, \mathbf{r}_2) = \Psi(\mathbf{r}_2, \mathbf{r}_1)$). Particles with half-integer spin are called *fermions* and particles with integer spin are called *bosons*.

The H nucleus has spin $\frac{1}{2}$ and is a fermion. The D nucleus has spin 1 and is a boson. HD does not contain identical nuclei and is not subject to these symmetry regulations. Moreover, the regulations for D_2 and H_2 are different, because D is a boson and H is a fermion. This might explain why Eq. 9.54 works for HD but not for D_2 and H_2, and why D_2 and H_2 misbehave differently.

In Supplement 9.1, I show that implementing the symmetry constraints does lead to new equations for the rotational energies that are in complete agreement with spectroscopic and thermodynamic measurements. Furthermore, in the limit of high temperature (i.e. $T \gg T_r$), those equations lead to Eq. 9.54 with $\sigma = 2$ for homonuclear molecules and $\sigma = 1$ for heteronuclear molecules.

This is an example that makes us believe that statistical mechanics and quantum mechanics are consistent and predictive theories. The symmetry principle was introduced to explain the properties of electrons in metals and molecules. Nobody imagined that it would deeply affect the calculation of the thermodynamic functions of gaseous H_2, and D_2. Nevertheless, when ignored, this principle leads to disagreement with experiment in a most unexpected way. When taken into account, the agreement is restored.

Furthermore, Dennison noted that the observed values for the C_p of H_2 agreed with the theory only if the spin of the proton was $\frac{1}{2}$. At that time the spin of the proton was not known; in fact no one knew whether protons had a spin. Later, direct measurements of the magnetic moment of the proton confirmed this prediction.

By using theories meant to explain the properties of electrons in metals and the photons emitted by a black body, we managed to explain (10 years later) the strange thermodynamics of H_2, HD, and D_2 and to "derive" the spin of the proton.

Later, this postulate explained perfectly the behavior of protons in molecular beam experiments measuring their magnetic moment. Those modern philosophers who argue that science is just a cultural convention, such as voodoo or the belief that the earth is flat, shared by a small number of people who read the same books, would perhaps reconsider their theories if they understood the facts mentioned above.

§17. *Rotational Contributions to Thermodynamic Quantities.* The rotational Helmholtz free energy per mole is

$$a_{\mathrm{r}} = -RT\ln\left(\frac{T}{\sigma T_{\mathrm{r}}}\right) \quad \text{for } T \gg T_{\mathrm{r}} \tag{9.55}$$

where $\sigma = 2$ for homonuclear diatomics (e.g. I_2) and $\sigma = 1$ for heteronuclear ones (e.g. HCl).

- The rotational contribution to energy is (use Eqs 9.55 and 1.9)

$$u_{\mathrm{r}} = -T^2\frac{\partial}{\partial T}\left(\frac{a_{\mathrm{r}}}{T}\right)_{V,N} = RT \quad \text{for } T \gg T_{\mathrm{r}} \tag{9.56}$$

- The rotational contribution to entropy is (use Eqs 9.55 and 1.6)

$$s_{\mathrm{r}} = -\left(\frac{\partial a_{\mathrm{r}}}{\partial T}\right)_{V,N} = R\ln\left(\frac{T}{\sigma T_{\mathrm{r}}}\right) + R \quad \text{for } T \gg T_{\mathrm{r}} \tag{9.57}$$

  Note that σ contributes to entropy but does not contribute to energy.

- The heat capacity at constant volume, is (use Eqs 9.55 and 1.11)

$$(C_v)_{\mathrm{r}} = \left(\frac{\partial u_{\mathrm{r}}}{\partial T}\right)_V = R \quad \text{for } T \gg T_{\mathrm{r}} \tag{9.58}$$

- The rotational contribution to chemical potential is (use Eqs 9.55 and 1.16)

$$\begin{aligned}\mu_{\mathrm{r}} &= \left(\frac{\partial A_{\mathrm{r}}}{\partial N}\right)_{T,V} = \frac{\partial}{\partial N}\left(-Nk_BT\ln\left(\frac{T}{\sigma T_{\mathrm{r}}}\right)\right) \\ &= -k_BT\ln\left(\frac{T}{\sigma T_{\mathrm{r}}}\right) \quad \text{for } T \gg T_{\mathrm{r}}\end{aligned} \tag{9.59}$$

This is the chemical potential per molecule. The chemical potential per mole is

$$\mu_r = -RT \ln \left(\frac{T}{\sigma T_r} \right) \quad \text{for } T \gg T_r \tag{9.60}$$

Exercise 9.15

Calculate $a_r(T) - a_r(298)$, $u_r(T) - u_r(298)$, $s_r(T)$, $\mu_r(T) - \mu_r(298)$, and $(C_v)_r$ for K_2 and H_2 at $p = 1$ bar and T varying from 400 K to 2000 K in increments of 200 K. Make a table with the results and compare H_2 and K_2. Also compare the rotational contribution to these quantities to the contribution made by vibrations and translations.

§18. *The Probability that a Molecule is Rotationally Excited.* The probability that a molecule is in the *rotational state j* is

$$p_j = \frac{\exp[-\varepsilon_j/k_B T]}{\sum_\ell \exp[-\varepsilon_\ell/k_B T]}$$

The sum in the denominator is q_r and $q_r = T_r/T$ if $T_r \ll T$ (see Eq. 9.51). Therefore, if $T/T_r \gg 1$, I have

$$p_j = \frac{T_r}{T} \exp\left[-\frac{j(j+1)B}{k_B T} \right] = \frac{T_r}{T} \exp\left[-\frac{j(j+1)T_r}{T} \right]$$

I used here the equation $\varepsilon_j = j(j+1)B$ and the notation $T_r = B/k_B$. Since, for most molecules at room temperature, $T_r \ll T$ (see Table 9.1), the exponential is large until $j(j+1)T_r/T \approx 2$. This means that all the states j that satisfy this condition are populated. For a molecule like I_2, for which $T_r = 0.0537$ K, it is quite likely to find molecules in the state $j = 100$ (see Fig. 9.2).

Because the rotational states are degenerate, the probability of finding a molecule having the *rotational energy* $\varepsilon_j = j(j+1)B$ is different from the probability of having a molecule in state j. Indeed, there are $2j+1$ states having the energy ε_j (which differ through the orientation of the angular momentum). Therefore the probability of finding a molecule with the energy ε_j is

$$p(\varepsilon_j) = \frac{(2j+1)\exp\left[-j(j+1)B/k_B T\right]}{q_r} = \frac{(2j+1)\exp\left[-j(j+1)T_r/T\right]}{q_r}$$

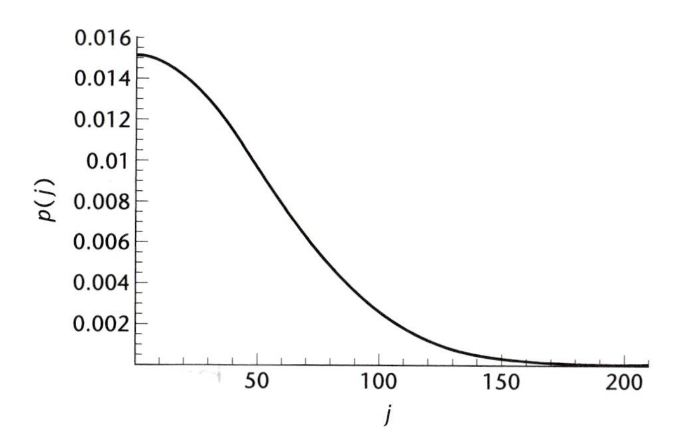

Figure 9.2 The probability of having a molecule in state j, as a function of j.

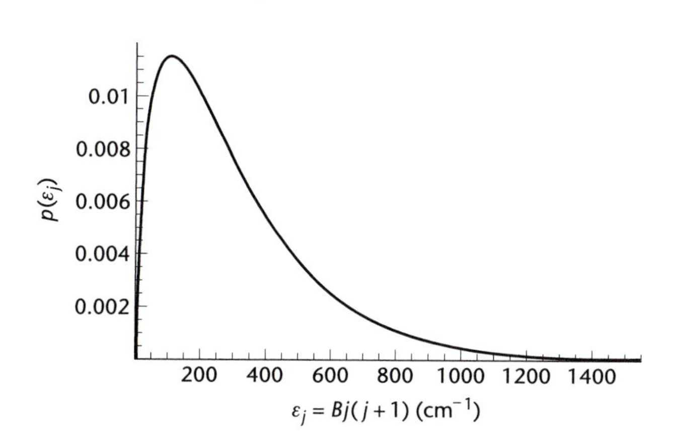

Figure 9.3 The probability of having a molecule with energy ε_j, as a function of the rotational energy.

This function looks quite different from $p(j)$, as you can see in Fig. 9.3. The calculation was performed in Workbook SM9.4.

§19. *Hot Bands in Spectroscopy.* The fact that in a gas of diatomic molecules so many rotational states are populated is a nightmare for spectroscopists. The molecules in different j states have slightly different absorption spectra. When one performs an absorption experiment, one measures all these spectra simultaneously: the molecule in $j = 0$ gives one spectrum, that in $j = 1$ gives another, etc. These additional spectra, caused by light absorption by molecules that are not on the ground state, are called *hot bands*.

The presence of the hot bands makes it hard to understand the spectrum. However, serendipity helps. If one passes the molecules through a narrow nozzle, from a high-pressure region into vacuum, the molecules emerging in the vacuum are in the ground vibrational and rotational states. All unwanted, hot spectra disappear.

Another method for getting rid of hot bands and cleaning up the absorption spectrum is to freeze the molecule in a solid matrix. Pinned in the solid, the molecule can no longer rotate and the hot spectra disappear.

§20. *The Contribution of the Nuclear Degrees of Freedom.* The contribution of the nuclear degrees of freedom to the free energy per mole is

$$a_{\mathrm{n}} = -RT \ln q_{\mathrm{n}} = -RT \ln \left([2I_1 + 1][2I_2 + 1]\right) \tag{9.61}$$

where I_α is the spin of nucleus α.

Using the equations given in Chapter 1 (they connect the thermodynamic functions to the Helmholtz free energy), we can easily derive the equations given below.

- The contribution to the energy of 1 mole of gas is

$$u_{\mathrm{n}} = RT^2 \frac{\partial}{\partial T}\left(\frac{a}{T}\right) = -RT^2 \frac{\partial}{\partial T}\left(R \ln \left([2I_1 + 1][2I_2 + 1]\right)\right) = 0 \tag{9.62}$$

- Since $u_{\mathrm{n}} = 0$, the contribution to heat capacity at constant volume is

$$(C_v)_{\mathrm{n}} = \frac{\partial u_{\mathrm{n}}}{\partial T} = 0 \tag{9.63}$$

- The contribution to the entropy is

$$s_{\mathrm{n}} = -\frac{\partial a_{\mathrm{n}}}{\partial T} = R \ln \left([2I_1 + 1][2I_2 + 1]\right) \tag{9.64}$$

- The contribution to the chemical potential per molecule is

$$\mu_{\mathrm{n}} = \frac{\partial a_{\mathrm{n}}}{\partial N} = \frac{\partial}{\partial N}\left(-k_B N T \ln q_{\mathrm{n}}\right) = -k_B T \ln \left([2I_1 + 1][2I_2 + 1]\right) \tag{9.65}$$

Note a general property: if a partition function is independent of temperature, it contributes to a, s, and μ, but not to u or C_v.

§21. *Put it all Together.* To obtain the expression for a given thermodynamic function I must add the contributions from the translational, rotational, vibrational, nuclear, and electronic degrees of freedom. Here I neglect the electronic contribution; this is valid for diatomics whose electronic excitation energies are much larger than $k_B T$.

- The Helmholtz free energy per mole is:

$$\begin{aligned} a &= a_\mathrm{t} + a_\mathrm{v} + a_\mathrm{r} + a_\mathrm{n} \\ &= -RT \ln\left(\frac{Ve}{N\Lambda^3}\right) - N_A D_0 + RT \ln\left(1 - \exp\left[-\frac{T_\mathrm{v}}{T}\right]\right) \\ &\quad - RT \ln\left(\frac{T}{\sigma T_\mathrm{r}}\right) - RT \ln\left([2I_1 + 1][2I_2 + 1]\right) \end{aligned} \tag{9.66}$$

To obtain the last equality, I have used Eqs 9.21, 9.39, 9.55, and 9.61.

- The total energy per mole is (use Eq. 6.23 with $N = N_A$ and $R = N_A k_b$) for the translational contribution, Eq. 9.40 for the vibrational contribution, and Eq. 9.56 for the rotational contribution).

$$u(T) = \frac{3}{2}RT - N_A D_0 + \frac{RT_\mathrm{v}}{\exp[T_\mathrm{v}/T] - 1} + RT \tag{9.67}$$

The nuclear partition function does not contribute to the energy of the gas since we have chosen the zero of energy to make the energy of the nucleus plus the energy of the electronic ground state equal to zero.

- The total entropy per mole $s(T, V)$ is given by

$$\begin{aligned} \frac{s(T,V)}{R} &= \frac{5}{2} + \ln\left(\frac{V}{N_A \Lambda(T)^3}\right) + \frac{T_\mathrm{v}/T}{\exp[T_\mathrm{v}/T] - 1} \\ &\quad - \ln\left(1 - \exp[-T_\mathrm{v}/T]\right) + R \ln\left([2I_1 + 1][2I_2 + 1]\right) \end{aligned} \tag{9.68}$$

- The heat capacity at constant volume is (use Eq. 6.46 for the translational contribution, Eq. 9.41 for the vibrational contribution, and Eq. 9.58 for the rotational contribution)

$$\frac{C_v(T)}{R} = \frac{3}{2} + \left(\frac{T_v}{T}\right)^2 \frac{\exp[T_v/T]}{\exp[T_v/T] - 1} + 1 \tag{9.69}$$

- It is now easy to calculate the heat capacity at constant pressure from $C_p = C_v + R$
- and the enthalpy from (use the definition $h = u + pv$ and the ideal gas law) $h = u + RT$
- The chemical potential of 1 mole of gas is (recall the definition $\mu = a + pv$) $\mu = a + pv = a + RT$.

  Using Eq. 9.66 for a leads to

$$\frac{\mu}{RT} = \ln\left(\frac{V}{N_A \Lambda(T)^3}\right) - \frac{D_0}{RT} + \ln\left(1 - \exp\left[-\frac{T_v}{T}\right]\right)$$
$$- \ln\left(\frac{T}{\sigma T_r}\right) - RT \ln\left([2I_1 + 1][2I_2 + 1]\right) \tag{9.70}$$

In these equations,

$$\Lambda(T) = \sqrt{\frac{h^2}{2\pi m k_B T}} \tag{9.71}$$

$$T_v = \frac{\hbar\omega}{k_B} \tag{9.72}$$

$$T_r = \frac{B}{k_B} \tag{9.73}$$

N_A is Avogadro's number, m is the mass of the diatomic, and I_1 and I_2 are the spins of the nuclei. Spectroscopic data (i.e. ω and B) can be found in the book by Huber and Herzberg cited on p. 126. The data from that book are reproduced at the NIST website. Data for thermodynamic functions are given at the same website.

To obtain these equations, I made several approximations: I used the harmonic approximation for vibrations; I assumed that rotations and vibrations are decoupled; I neglected the electronic degrees of freedom; I assumed that the gas is ideal. I also used a simplified formula for rotations, which works well if $T \gg T_r$, but gives erroneous results for H_2 and D_2 at low temperatures ($T < 200$ K). The rotational temperatures of most of the other diatomics are so small that $T \gg T_r$ at the temperatures at which they are still gases. Therefore, the equations derived here for rotational partition function work for them.

At low temperatures and/or high pressures the assumption of ideality breaks down. At very high temperatures the harmonic approximation and the decoupling of rotation from vibration break down. In the next chapter, these equations are tested by comparison with experiment.

Supplement 9.1 The Rotational Partition Function for Homonuclear Diatomic Molecules

§22. *Introduction.* Earlier in this Chapter I have shown that at high temperatures, the rotational partition function of a diatomic molecule is

$$q_r = \frac{T}{\sigma T_r} \tag{9.74}$$

Here T_r is the rotational temperature and σ is the mysterious *symmetry number*, which is equal to 1 if the molecule is heteronuclear (different nuclei) and equal to 2 if it is homonuclear (identical nuclei). When I introduced this equation, I warned you that it is correct only if

$$T \gg T_r \tag{9.75}$$

Eq. 9.74 was derived by taking advantage of the fact that when $T \gg T_r$ we can turn the sum

$$q_r = \frac{1}{\sigma} \sum_{j=0}^{\infty} (2j+1) \exp\left[-\frac{j(j+1)T_r}{T}\right] \tag{9.76}$$

into an integral; performing the integration led to Eq. 9.74. With today's computers we can evaluate the sum and remove this particular approximation. Unfortunately, using the sum instead of the integral still leads to discrepancies with experiment.

The inadequacy of the theory presented above was discovered while comparing the heat capacities of H_2, D_2 calculated by using Eq. 9.76 for the rotational partition function, with the values measured at low temperatures. H_2, D_2, and

HD played a prominent role because they have the highest rotational temperature T_r and become liquid at very low temperatures. It is therefore possible to make measurements of the heat capacity of the gas at temperatures comparable to or lower than T_r. It is in this regime that data show that a theory based on Eq. 9.76 falls apart.

The mystery deepened when it was discovered that the heat capacity of HD agrees well with the low-temperature data, as long as we calculate it by using Eq. 9.76 and not Eq. 9.74.

Try to imagine how puzzled the people studying this problem in the 1920s must have been. H_2, D_2, and HD differ only through the number of *neutrons* in their nuclei. Given the knowledge available at that time, chemists were justified in believing that adding an *electrically neutral* particle to a nucleus in a molecule can change no other property but the mass. But here it was: a theory that took this mass change correctly into account led to a heat capacity that disagreed dramatically with the facts. It must have been particularly intriguing to see that a theory that works for HD fails for H_2 and D_2. This suggested that, perhaps, the theory is deficient only when the molecule has identical nuclei.

Exercise 9.16

Analyze the partition function of the diatomic molecule and explain which quantities change with the nuclear mass and how.

In this supplement I show that the error in the theory appears because we have ignored the requirements of one of the fundamental principles of quantum mechanics: the wave function of a molecule must change in a certain way when we permute the coordinates of the identical nuclei.

Supplement 9.2 Symmetry Requirements

§23. *The Total Wave Function must be Symmetric or Antisymmetric.* When you studied quantum mechanics you learned that the wave function $\Psi(\mathbf{r}_1, \mathbf{r}_2)$ of a system of *identical particles* must be

- *antisymmetric* if the particles are *fermions* (i.e. they have half-integer spin, $\frac{1}{2}, \frac{3}{2}, \frac{5}{2}, \ldots$):

$$\Psi(\mathbf{r}_1, \mathbf{r}_2) = -\Psi(\mathbf{r}_2, \mathbf{r}_1) \tag{9.77}$$

Table 9.2 The spin I of a few nuclei, the number n_s of symmetric nuclear spin wave functions, and the number n_a of antisymmetric nuclear spin wave functions.

Nucleus	I	n_s	n_a
^{1}H	$\frac{1}{2}$	3	1
$^{2}H = D$	1	6	3
^{16}O	0	1	0
^{17}O	$\frac{5}{2}$	21	15
^{18}O	0	1	0

and

- *symmetric* if the particles are *bosons* (i.e. they have integer spin, $0, 1, 2, \ldots$):

$$\Psi(\mathbf{r}_1, \mathbf{r}_2) = +\Psi(\mathbf{r}_2, \mathbf{r}_1) \tag{9.78}$$

Here $\mathbf{r}_1$ and $\mathbf{r}_2$ are the positions of the nuclei.

Since the spin of the proton is $I = \frac{1}{2}$ (see Table 9.2), the wave function of the H_2 molecule must be *antisymmetric* when we permute $\mathbf{r}_1$ with $\mathbf{r}_2$ (it must satisfy Eq. 9.77).

The spin of the deuteron is $I = 1$ and the wave function of the D_2 molecule must be *symmetric* with respect to a permutation of $\mathbf{r}_1$ and $\mathbf{r}_2$ (it must satisfy Eq. 9.78).

There are no symmetry requirements for HD since the two nuclei are not identical.

As you will see in what follows, these demands force us to modify the nuclear and rotational partition functions of homonuclear molecules.

I emphasize that the statement that two particles are identical is very strict. The molecule $^{17}O^{16}O$ is not homonuclear since it contains two different isotopes of the oxygen nucleus. On the other hand $^{16}O^{16}O$ is homonuclear.

§24. *The Symmetry of the Total Wave Function of a Homonuclear Diatomic Molecule.* The symmetry requirements must be satisfied by the *total* wave function $\Psi(\mathbf{r}_1, \mathbf{r}_2)$ of the diatomic. This is given by the product

$$\Psi(\mathbf{r}_1, \mathbf{r}_2) = \psi^{\mathrm{e}}(\mathbf{r}_1, \mathbf{r}_2)\, \psi^{\mathrm{n}}\, \psi_i^{\mathrm{v}}(r)\, \psi_j^{\mathrm{r}}(\theta, \phi) \tag{9.79}$$

Here $\psi^{\mathrm{e}}(\mathbf{r}_1, \mathbf{r}_2)$ is the wave function of the electrons in the H_2 molecule, ψ^{n} is the wave function of the nuclear spins, $\psi_i^{\mathrm{v}}(r)$ is the vibrational wave function, and $\psi_j^{\mathrm{r}}(\theta, \phi)$ is the rotational wave function. The positions of the nuclei are given by the vectors $\mathbf{r}_1$ and $\mathbf{r}_2$, r is the distance between the nuclei, and θ and ϕ are the polar and azimuthal angles of the molecular axis.

If the two identical nuclei are fermions, permuting $\mathbf{r}_1$ with $\mathbf{r}_2$ must cause $\Psi(\mathbf{r}_1, \mathbf{r}_2)$ to change sign. If the identical nuclei are bosons, the function must remain unchanged when we interchange $\mathbf{r}_1$ with $\mathbf{r}_2$.

To understand how $\Psi(\mathbf{r}_1, \mathbf{r}_2)$ changes when we interchange $\mathbf{r}_1$ with $\mathbf{r}_2$, we must find out how every function present in the product in Eq. 9.79 changes upon this permutation.

The electronic wave function $\psi^{\mathrm{e}}(\mathbf{r}_1, \mathbf{r}_2)$ does not change when we interchange the coordinates of the two nuclei.

The vibrational wave function $\psi_i^{\mathrm{v}}(r)$ depends only on the distance between the nuclei. This distance remains *unchanged* when we interchange $\mathbf{r}_1$ with $\mathbf{r}_2$ and therefore the vibrational wave function will not change.

The rotational wave function $\psi_j^{\mathrm{r}}(\theta, \phi)$ is more interesting.

- If j *is odd*, then $\psi_j^{\mathrm{r}}(\theta, \phi)$ changes sign when the positions $\mathbf{r}_1$ and $\mathbf{r}_2$ of the nuclei are interchanged (the function is *antisymmetric*).
- If j *is even*, then $\psi_j^{\mathrm{r}}(\theta, \phi)$ remains unchanged when the positions of the nuclei are interchanged (the function is *symmetric*).

Exercise 9.17

If you review what you have learned about the rigid rotor in quantum mechanics, you can check that the statements above are true, by exchanging the nuclei in the rotational wave functions (the spherical harmonic $Y_j^m(\theta, \phi)$).

§25. *The Role of Nuclear Spin.* Finally, we have to consider the nuclear spin wave function. As emphasized, the energy required to excite a nucleus is so large that we only need to include the contribution from the ground state.

An unexcited nucleus can have several nuclear states, which differ through their spin, and whose energies are very close to each other.

§26. *Spin One-Half.* As a first pass at understanding what these states are, I will remind you what you have learned when you studied the quantum mechanics of the electrons in the H_2 molecule. When you constructed the wave function of the two electrons you had to consider the spin wave function along with the molecular orbitals. The electrons have spin $\frac{1}{2}$ and the two electrons could be in

- either three symmetric spin states (triplet)

$$\frac{1}{\sqrt{2}}[\alpha(1)\beta(2) + \alpha(2)\beta(1)]$$

$$\alpha(1)\alpha(2)$$

$$\beta(1)\beta(2)$$

- or one antisymmetric wave function (singlet)

$$\frac{1}{\sqrt{2}}[\alpha(1)\beta(2) - \alpha(2)\beta(1)]$$

Here α is a state with "spin up" and β has "spin down" (1 and 2 label the electrons).

This analysis is *valid for any particle that has spin* $\frac{1}{2}$. Since the spin of the proton is $\frac{1}{2}$, the same statements hold for the spin states of the two nuclei in H_2.

In what follows I will call an H_2 molecule having a symmetric nuclear-spin state *orthohydrogen* and one in an antisymmetric state *parahydrogen*.

The energy of E_p of the nuclear spins in parahydrogen is slightly lower than the energy E_o of nuclear spins in orthohydrogen. The difference $E_o - E_p$ is smaller that $k_B T$ at all but the lowest temperatures. Because of this, I will assume that $E_o - E_p \approx 0$. This allows me to set $\exp\left[-(E_o - E_p)/k_b T\right] \approx 1$ when I calculate the nuclear partition function of the molecule. Calculations at very low temperatures will have to take such terms into account.

Exercise 9.18

When you read what follows put the terms $\exp\left[-E_o/k_BT\right]$ and $\exp\left[-E_p/k_BT\right]$ where they are needed (I will set them all equal to 1). Then argue that the partition function of the nuclear spins depends only on $\exp\left[-(E_o - E_p)/k_BT\right]$. Discuss how the population of the ortho- and parahydrogen changes with temperature, when the system is in equilibrium. To perform numerical calculations of the population, use $E_o - E_p = 453$ J/g (convert this energy into energy per molecule).

§27. *Fermions with Arbitrary Spin.* The spin I of many nuclei differs from $\frac{1}{2}$. The theory of spin (see any book of quantum mechanics) tells us that, regardless of whether the nuclei are bosons or fermions, two nuclei, each having spin I, have

- either

$$I(2I + 1) \tag{9.80}$$

  antisymmetric spin states

- or

$$(I + 1)(2I + 1) \tag{9.81}$$

  symmetric states.

The energy differences between these states are small compared to k_BT and for simplicity I assume that they all have the same energy. This means that when I calculate the nuclear partition functions I take all the terms of the form $\exp\left[-E/k_BT\right]$, where E is the energy of the nuclear spin state, to be equal to 1.

Exercise 9.19

Show that Eqs 9.80 and 9.81 give correct results for H_2.

Exercise 9.20

Write the expression for the nuclear partition function for a homonuclear diatomic molecule whose nuclei have the spin one. Do not ignore the fact that the spin states have slightly different energies. Use the symbols E_1, E_2, etc. for these unknown energies.

§28. *Summary*. This ends the enumeration of the symmetry properties of the wave functions whose product gives the total wave function. The electronic and the vibrational wave functions do not change when we permute the position of the two nuclei. Therefore, to impose the correct symmetry on the wave function of the molecule, we only need to consider the symmetry of the nuclear and rotational states. For this reason the symmetry requirements will affect only the rotational and the nuclear partition functions.

Supplement 9.3 The Nuclear-Rotational Partition Function of a Homonuclear Diatomic Whose Nuclei are Fermions

Let us begin with the case when the two nuclei are fermions. The total wave function must be antisymmetric. This can happen in two ways: either (1) the nuclear-spin wave function is symmetric and the rotational wave function is antisymmetric or (2) the spin function is antisymmetric and the rotational function is symmetric. No other combination is allowed by the symmetry requirements.

§29. *Symmetric Nuclear-Spin States.* If the nuclear-spin wave function ψ^{n} is symmetric, the rotational one ψ_j^{r} must be antisymmetric. This means that the molecule *can only have rotational states with odd values of* j (see §24). Since the nuclear state is symmetric, the nuclear degeneracy is $(I+1)(2I+1)$ (see §27). The contribution of these states to the rotational-nuclear partition function is therefore

$$q_{rn}^{o} = (I+1)(2I+1) \sum_{j \text{ odd}}^{\infty} (2j+1) \exp\left[-\frac{j(j+1)T_{\mathrm{r}}}{T}\right] \tag{9.82}$$

The sum is performed only over odd values of the rotational quantum number j. When we apply this equation to H_2 we obtain the partition function for orthohydrogen. We can generalize this nomenclature and use the prefix "ortho" for any homonuclear diatomic that is in a symmetric spin state. This is why I used the superscript o in Eq. 9.82.

§30. *Antisymmetric Nuclear-Spin States.* If the nuclear spin function is antisymmetric, the rotational wave function must be symmetric (so that their product is antisymmetric). The rotational wave function is symmetric only if *the rotational quantum number j is even* (see §24). Because of the symmetry rules, the molecule cannot have states with odd values of j. The nuclear degeneracy of the antisymmetric nuclear-spin states is $I(2I+1)$ (see §27). The contribution of these states to the rotational-nuclear partition function is

$$q_{rn}^{p} = I(2I+1) \sum_{j \text{ even}}^{\infty} (2j+1) \exp\left[-\frac{j(j+1)T_{\mathrm{r}}}{T}\right] \tag{9.83}$$

The superscript p indicates that this is the contribution of the antisymmetric nuclear-spin states to the partition function. In the case of H_2, this is the partition function of parahydrogen.

§31. *The Total Nuclear-Rotational Partition Function of a Homonuclear Diatomic whose Nuclei are Fermions.* To obtain the partition function q_{rn} of the rotational and nuclear degrees of freedom, we must add the contributions from all possible states. This means that q_{rn} is the sum of Eqs 9.82 and 9.83:

$$q_{\mathrm{rn}} = (I+1)(2I+1) \sum_{j \text{ odd}}^{\infty} (2j+1) \exp\left[-\frac{j(j+1)T_{\mathrm{r}}}{T}\right] + I(2I+1) \sum_{j \text{ even}}^{\infty} (2j+1) \exp\left[-\frac{j(j+1)T_{\mathrm{r}}}{T}\right] \tag{9.84}$$

This nuclear-rotational partition function must be valid for all homonuclear diatomics with half-integer spins (e.g. H_2, $^{17}O^{17}O$). In particular, it must become equal to $(2I+1)^2 T_{\mathrm{r}}/2T$ when $T \gg T_{\mathrm{r}}$ (since this equation agrees with experiments in that temperature range).

You should remember that in deriving this result we have chosen to ignore the small differences in the energies of different nuclear spin states. This is a good approximation at the temperatures of interest to us.

Exercise 9.21

Denote the energies of the nuclear spin states by E_1, E_2, ... and put the exponential terms $\exp[-E_i/k_B T]$ where they belong, in the equations obtained above.

§32. *Comment.* The partition function q_{rn} no longer has the form $q_{\mathrm{r}}q_{\mathrm{n}}$. This is an odd situation: I have stated earlier in this chapter that the energy of the molecule is a sum of nuclear, rotational, vibrational, electronic, and translational energies. I made this statement because the Hamiltonian of the molecule was a sum of terms, one term for each degree of freedom. In this case we can prove mathematically, with no approximations, that the energy is the sum mentioned above. Then, pure mathematics leads us to conclude that the partition function of a molecule is a product of the partition function of the degrees of freedom. Now I am telling you that this argument is erroneous, as far as the nuclear and rotational degrees of freedom are concerned.

What went wrong in the above argument? The error crept in when we analyzed the Schrödinger equation without imposing the antisymmetry condition (for fermions) or the symmetry condition (for bosons). Once we impose these, we find that the symmetry of the nuclear-spin state affects the rotational energies. If the nuclei are fermions and the nuclear-spin state is antisymmetric, then only rotational states with even j are permitted. The states with odd j, which are allowed when symmetry requirements are ignored, are no longer permitted! Clearly the nuclear state affects the rotational state, even though there is no interaction between them in the Hamiltonian. Because of the symmetry constraints, two degrees of freedom are no longer independent and the partition function is no longer a product.

Exercise 9.22

What are consequences of the restrictions imposed by the symmetry requirements for the spectroscopy of homonuclear diatomic molecules?

A similar "interaction" appeared when we divided the partition function of the gas by $N!$. This factor was required by the same symmetry conditions we used here and the outcome was that a new term appeared in the partition function of the gas that was equivalent to an interaction between the molecules. This interaction was present even though we used a Hamiltonian in which the molecules did not interact.

§33. *High-Temperature Limit.* We know that when $T \gg T_{\mathrm{r}}$ the equation

$$q_{\mathrm{rn}} = (2I+1)^2 \frac{T_{\mathrm{r}}}{2T} \tag{9.85}$$

agrees with experiment. The factor of 2 is the famous symmetry number, which is causing us so much headache. We introduced it without a sound theoretical reason. If our symmetry-based analysis is correct, the high-temperature limit of Eq. 9.84 should give Eq. 9.85; there should be no need to introduce a factor of 2 arbitrarily.

Consider what happens to the two sums present in Eq. 9.84 when $T \gg T_r$. The function

$$f(j) = (2j+1)\exp\left[-j(j+1)\frac{T_r}{T}\right]$$

changes very little when we change j (if $T \gg T_r$). Because of this

$$\sum_{j\geq 0,\text{ even}} f(j) \approx \sum_{j\geq 1,\text{ odd}} f(j) \approx \frac{1}{2}\sum_{j=0}^{\infty} f(j) \tag{9.86}$$

Exercise 9.23

Write a computer program that calculates the three expressions present in Eq. 9.86 and show that they are equal when $T \gg T_r$. Determine at what temperatures these expressions are no longer equal.

But

$$\frac{1}{2}\sum_{j=0}^{\infty} f(j) = \frac{1}{2}\sum_{j=0}^{\infty}(2j+1)\exp\left[-\frac{j(j+1)T_r}{T}\right] = \frac{T}{2T_r} \tag{9.87}$$

The last equality was proven in §15.

This calculation shows that the sums over even or odd j, which appear in Eq. 9.84, are equal to $T/2T_r$. Using this result in Eq. 9.84 leads to

$$\begin{aligned} q_{rn} &= [(I+1)(2I+1) + I(2I+1)]\frac{T_r}{2T} \\ &= (2I+1)^2\frac{T_r}{2T} \quad \text{when } T \gg T_r \end{aligned} \tag{9.88}$$

This is Eq. 9.85, derived by ignoring the symmetry requirements. The factor of 2 appears naturally when the symmetry conditions required for dealing with identical particles are taken into account. This derivation of the symmetry number has been done for fermions. The same result is obtained for bosons.

---

§34. *The Temperature Dependence of C_v for H_2.* At temperatures comparable to or lower than T_r we must use Eq. 9.84 to calculate thermodynamic quantities. Here I calculate C_v for H_2. This is given by

$$C_v = (C_v)_t + (C_v)_v + (C_v)_{rn} + (C_v)_e \tag{9.89}$$

The translational contribution $(C_v)_t$ is (see Eq. 6.53)

$$(C_v)_t = \frac{3R}{2} \tag{9.90}$$

The vibrational contribution $(C_v)_v$ is (see Eq. 9.41)

$$(C_v)_v = R\left(\frac{T_v}{T}\right)^2 \frac{\exp\left[\frac{T_v}{T}\right]}{\left(\exp\left[\frac{T_v}{T}\right] - 1\right)^2} \tag{9.91}$$

Since H_2, D_2 and HD have very high vibrational temperatures, the vibrational motion makes a very small contribution to C_v. Nevertheless, since the computer does all the work, I include the vibrational contribution.

The rotational-nuclear contributions must be treated with care. The heat capacity is calculated from (see Eq. 1.11):

$$(C_v)_{rn} = \frac{\partial u_{rn}}{\partial T} \tag{9.92}$$

To obtain the energy u_{rn}, I use Eq. 1.9

$$u_{rn} = -T^2 \frac{\partial}{\partial T}\left(\frac{a_{rn}}{T}\right)_{V,N} \tag{9.93}$$

with $a_{rn} = -RT \ln\left(q_{rn}\right)$ and q_{rn} calculated with Eq. 9.84.

§35. *Does Taking Symmetry into Account Make a Substantial Difference in the Heat Capacity?* The first question we ask is whether taking the symmetry into account makes a difference in the dependence of C_v on temperature. In Workbook SM9.5 I calculate C_v with the symmetry restriction and without. The calculation poses no special difficulty: I defined functions that calculate the partition functions and programmed the computer to perform the derivatives required by the equations above. I used the following data for H_2: the vibrational energy is $\hbar\omega = 4401\ \text{cm}^{-1}$, the rotational constant is $B = 60.853\ \text{cm}^{-1}$, and Boltzmann constant is $k_B = 0.695\ \text{cm}^{-1}/\text{K}$.

In Fig. 9.4 I show the heat capacity C_v for H_2 calculated correctly (dashed line) and also calculated by ignoring the requirements of symmetry (solid line). Imposing the rule that the wave function must be antisymmetric does make a big difference — not just in the numerical values: the shape of the curve is different.

§36. *Does this Theory Agree with the Experiments?* The next, and more important, question is whether the corrected theory agrees with experiment. The short answer is no. The history of the subject is rather interesting and somewhat contorted, and it is instructive to review it.

Pauli introduced the symmetry principle in a crude way, by stating that no two electrons in a molecule can have the same state. This was greatly generalized

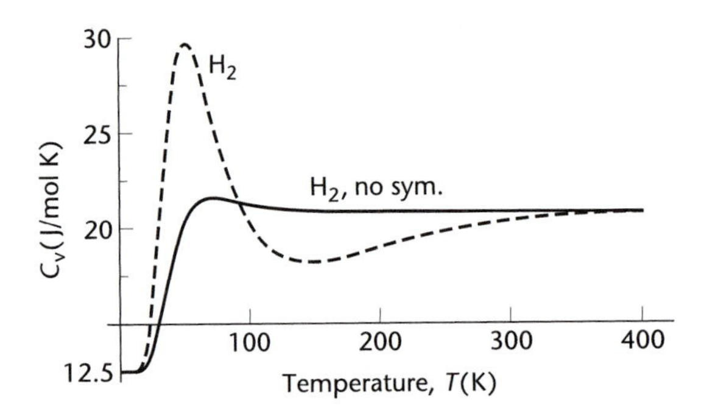

Figure 9.4 The temperature dependence of heat capacity at constant volume, C_v, of H_2 at low temperatures. The solid line shows the heat capacity calculated by ignoring the symmetry requirements (Eq. 9.76) for q_{rn}, and the dotted line shows the correct calculation (using Eq. 9.84 for the partition function q_{rn}).

by Dirac who formulated the principle in the form used today. The inventor of quantum mechanics, Werner Heisenberg[e], was the first to suggest that the H_2 molecule comes in two forms, orthohydrogen and parahydrogen, having different spin states of the nuclei. At that time it was not known that protons have spin $\frac{1}{2}$, so treating the proton as a fermion was a tentative (and inspired) guess.

Hundt[f] realized that if ortho- and parahydrogen existed, they will affect the rotational and the nuclear partition function of the molecule. He proposed replacing $T/2T_r$ with Eq. 9.84. He used this partition function to calculate C_v, but proceeded a bit differently from us. He did not know the molecular constants for H_2, so he adjusted them to fit the experiments. The fit was rather good and Hundt thought that he had solved the problem.

But this was not the end of the story. In the same year, Dennison[g], who was a spectroscopist, noticed that the molecular constants obtained by Hundt, when he fitted the heat capacity data, were not in agreement with those measured by spectroscopy. He concluded, correctly, that agreement with experiment found by Hundt was accidental. This sort of fake agreement is not as rare as you might think, when parameters in an equation are varied to fit the data.

Dennison solved the impasse by proposing an ingenious solution. To understand his reasoning we need to learn more about ortho- and parahydrogen.

§37. *Orthohydrogen and Parahydrogen.* Let us assume that somehow we have prepared a gas of pure parahydrogen. What happens to this gas after preparation? We would expect the gas to go to equilibrium. Since orthohydrogen has three nuclear-spin states and parahydrogen has one, we expect that at equilibrium there will be three orthohydrogen molecules for each parahydrogen molecule. (I have assumed that these states have the same energy and I have taken it to be zero.) There is no doubt that sooner or later the equilibrium state is reached. What is important is to know how quickly.

To answer that we need to examine the process that causes a molecule to change its nuclear-spin state from a singlet to a triplet. The molecules in a gas change their state because of collisions with other molecules. To change

[e] W. Heisenberg, *Zeit. f. Physik* **41**, 239 (1927).

[f] F. Hundt, *Zeit. f. Physik* **42**, 93 (1927).

[g] D.M. Dennison, *Proc. Roy. Soc. A* **115**, 483 (1927).

the nuclear-spin state, some sort of magnetic force (spins interact with magnetic fields only) must act on the nuclear spin during the collision. It so happens that the magnetic field exerted by one H_2 molecule on its collision partner is extremely small; therefore the nuclear spin is very rarely affected by a collision. The collision may be very violent and cause a great change in the rotational and vibrational energy of the colliding molecules, but the nuclear spins remain unchanged, oblivious to the trauma experienced by the other degrees of freedom.

If you think that this strains belief consider two other facts. Nuclear physicists like to study the properties of bare nuclei (those with no electrons around them) that have a specific nuclear spin (they call it a spin-polarized nucleus). To prepare a spin-polarized nucleus they use laser spectroscopy to create ions whose nucleus is spin-polarized. To obtain the bare nucleus they need to strip off all the electrons from the ion, without disturbing the spin of the nucleus. This is how they do it. They accelerate the ion (with electric fields) to enormous energies and pass it through a very thin metal foil. All kinds of events take place during the passage through the foil. The particles that pass through the foil are very highly ionized atoms and bare (no electrons) nuclei. The spin state of practically all bare nuclei prepared in this way is the same as it was when the ion approached the foil. In spite of the extraordinary destruction that the ion suffers, the nuclear spin goes through the solid as if nothing happens.

How it this possible? The electrons are stripped because enormous electrostatic forces act on them as they pass through the foil. However, the spin is affected only by magnetic forces and these happen to be extremely small.

In a different kind of experiment Professor Alex Pines, at UC Berkeley, makes (by optical methods) Xe atoms with polarized nuclear spin. The gas is prepared in one building and then it is carried to a building on the hill overlooking the campus, where it is used in a variety of experiments. After this fairly long trip, the nuclear spins are still polarized.

None of these facts were known to Dennison, but he knew that the magnetic fields acting on the nuclei during collisions were very small, which meant that the collisions between the molecules will change the spin state of their nuclei very rarely. He assumed that the time in which the nuclear spin changes its state is much longer than the time during which the heat capacity is measured. In this case we will have to think of the gas as a mixture of ortho- and parahydrogen. The rotational-nuclear contribution to the heat capacity of this mixture, $(C_v)_{n,r}$, should then be given by

$$(C_v)_{\text{nr}} = (C_v)^o_{\text{nr}} f_o + (C_v)^p_{\text{nr}} f_p \tag{9.94}$$

Here $(C_v)^o_{\text{nr}}$ and $(C_v)^p_{\text{nr}}$ are the heat capacities of the orthohydrogen and parahydrogen, respectively, and f_o and f_p are the fractions of the two components in the gas. To calculate $(C_v)^o_{\text{nr}}$, Dennison used the partition function q^o_{nr} of the orthohydrogen (Eq. 9.82) and for $(C_v)^o_{\text{nr}}$ he used the partition function q^p_{nr} of the parahydrogen (Eq. 9.82).

Since he did not know the fractions of ortho- and parahydrogen in the gas used for measurements, he varied f_o to fit the data (by definition $f_p = 1 - f_o$). The other quantities needed in the calculation were taken from spectroscopy. The results were in perfect agreement with the experiments.

The fraction of orthohydrogen that gave the best fit to experiments was $f_o = \frac{3}{4}$, which leads to $f_p = \frac{1}{4}$. There are three times as many orthohydrogen molecules as parahydrogen. This is what one would expect if the two species were in equilibrium, *at room temperature* (the orthohydrogen spin-state is triply degenerate and that of the parahydrogen is a singlet). The gas was then cooled and its heat capacity was measured. The fraction of the two components did not change during cooling: this is why $f_o = \frac{3}{4}$.

It is now possible to prepare mixtures of ortho- and parahydrogen of known concentrations, calculate the heat capacity by the procedure above, and find perfect agreement with the experiment without varying f_o to fit the data.

§38. *What is this good for?* If you are inclined to think that science is interesting only when it leads to applications, you may wonder why anyone would need to know the thermodynamic functions of hydrogen at such low temperatures. NASA uses liquid hydrogen as rocket fuel. At very low temperatures (hydrogen becomes liquid at 20 K), the equilibrium concentration of parahydrogen is 99.8%. However, before cooling the gas the fraction of orthohydrogen is much higher. After cooling to liquefy the hydrogen, the orthohydrogen changes very slowly into parahydrogen, to reach the equilibrium composition. The heat of this "reaction" is 532 J/g and it is greater than the heat of evaporation of the liquid parahydrogen (which is 453 J/g). The slow but inevitable conversion of orthohydrogen to parahydrogen causes the evaporation of the liquid parahydrogen, leading to unacceptable losses of fuel (about 1% of the fuel is lost per hour). These losses can be avoided by using parahydrogen as fuel.

Supplement 9.4 The Nuclear-Rotational Partition Function of a Homonuclear Diatomic Whose Nuclei are Bosons

§39. *Symmetry Requirements for Bosons.* It is straightforward to perform for bosons the type of calculations done for fermions. The total wave function of the molecule must be symmetric with respect to the permutation of the coordinates of the identical nuclei. As in the case of fermions we only need to examine the product of the rotational and nuclear wave functions, ψ^{r} and ψ^{n}.

- If ψ^{n} is symmetric, then the product $\psi^{\mathrm{r}}\psi^{\mathrm{n}}$ is symmetric only if ψ^{r} is symmetric. This means that j can only have even values. The degeneracy of the symmetric nuclear-spin wave function ψ^{n} is $(I+1)(2I+1)$ (see Eq. 9.81). The contribution of these states to the partition function q_{rn} is:

$$q_{rn}^{p} = (I+1)(2I+1)\sum_{j\ \mathrm{even}}(2j+1)\exp\left[-\frac{j(j+1)T_{\mathrm{r}}}{T}\right] \tag{9.95}$$

  We will use the prefix 'para' for the molecules that have a symmetric nuclear-spin wave function.

- We also obtain a symmetric total wave function if the nuclear and the rotational wave functions are both antisymmetric (then the product $\psi^{\mathrm{r}}\psi^{\mathrm{n}}$ is symmetric). This means that only odd values of j are allowed and that the nuclear degeneracy is $I(2I+1)$. These states contribute

$$q_{\mathrm{rn}}^{o} = I(2I+1)\sum_{j\ \mathrm{odd}}(2j+1)\exp\left[-\frac{j(j+1)T_{\mathrm{r}}}{T}\right] \tag{9.96}$$

  to the nuclear-rotational partition function.

The total partition function for bosons is the sum of Eq. 9.95 and Eq. 9.96:

$$q_{\mathrm{rn}} = (I+1)(2I+1)\sum_{j\ \mathrm{even}}(2j+1)\exp\left[-\frac{j(j+1)T_{\mathrm{r}}}{T}\right] + I(2I+1)\sum_{j\ \mathrm{odd}}(2j+1)\exp\left[-\frac{j(j+1)T_{\mathrm{r}}}{T}\right] \tag{9.97}$$

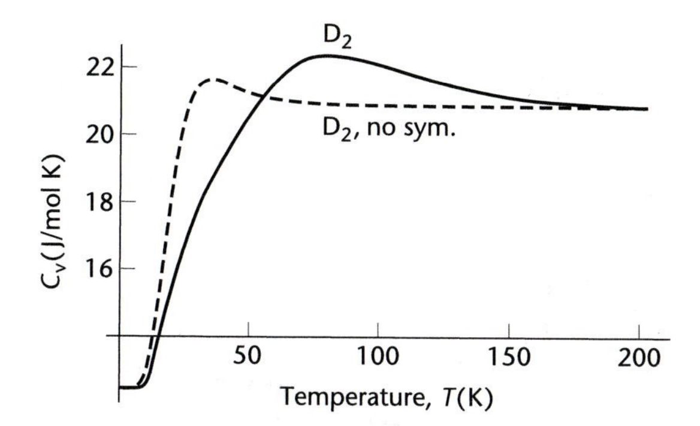

Figure 9.5 The temperature dependence of heat capacity at constant volume, C_V, of D_2 at low temperatures. The solid line shows the heat capacity calculated by ignoring the symmetry requirements (Eq. 9.76) for q_{rn}, and the dotted line shows the correct calculation (using Eq. 9.84 for the partition function q_{rn}).

However, because the conversion from a symmetric to an antisymmetric spin-state is very slow, we do not use this partition function to calculate thermodynamic quantities. It is more appropriate to consider that the system is a mixture of molecules with symmetric nuclear-spin states and molecules with antisymmetric nuclear-spin states. To perform such calculation we need to know the concentration of these two kinds of molecules in the mixture.

Exercise 9.24

Show that if $T \gg T_r$, Eq. 9.97 reduces to $q_{rn} = (2I + 1)^2 T_r/2T$.

Exercise 9.25

Plot the temperature dependence of C_v for D_2 versus temperature calculated by using Eq. 9.97 for q_{rn} and also by using the equation $q_{rn} = (2I + 1)^2 T_r/2T$. Is the effect of symmetry requirements important? Use $\hbar\omega = 3115.50\ \text{cm}^{-1}$ and $B = 30.443\ \text{cm}^{-1}$. See if you get the curve shown in Fig. 9.5.

10

A GAS OF DIATOMIC MOLECULES: COMPARISON WITH EXPERIMENT

Introduction

§1. *What to Look For.* In the previous chapter, I presented a theory that claims that, given molecular data, it can calculate all thermodynamic functions of an ideal gas composed of diatomic molecules. In this chapter, I will test this claim. To do this, I calculate the heat capacity C_p of several molecules and the entropy s of O_2, for various temperatures, at the pressure $p = 1$ bar. Then I compare the results to experiment.

But there is more to life than numbers. Look at Table 10.1, which gives the heat capacities of several molecules at several temperatures. The heat capacities of O_2, N_2, H_2, HCl, and CO are very close to each other. You might be ready to conclude that the C_p of every diatomic is roughly 29.2 J/mol K. However, this is not correct: the C_ps of K_2, Br_2, I_2, and Cl_2 are close to each other, but are substantially larger than those of O_2, N_2, H_2, HCl, and CO. Can we understand why there are similarities within a group and what causes the difference between the molecules in the two groups?

Table 10.1 The heat capacities (in J/mol K) at constant pressure of several diatomic gases. The data are from *NIST-JANAF Thermochemical Tables,* Fourth Edition, M. W. Chase, Jr., J. Phys. Chem. Ref. Data, Monograph 9 (1998), reproduced at webbook.nist.gov.chemistry. The right-hand column gives the vibrational temperature T_v.

Gas	Temperature, T(K)			T_v (K)
	298	500	1000	
CO	29.142	29.974	33.183	3121.48
N_2	29.124	29.580	32.690	3392.01
O_2	29.376	31.091	34.070	2273.64
H_2	28.836	29.260	30.025	6338.20
HCl	29.136	29.308	31.628	4301.38
Cl_2	33.949	36.064	37.438	807.30
I_2	36.887	37.464	38.081	308.65
K_2	37.980	38.750	36.250	133.00

In Chapter 3, we spent quite a bit of time developing some intuition regarding what makes entropy small or large. Now I can put this intuition to the test. For example, I will try to explain why the entropy of H_2 is smaller than that of K_2.

The equations needed for the calculations performed in this chapter were collected at the end of Chapter 9. The theory must agree with experiment as long as the gas is ideal. However, I expect some errors at high temperature because the harmonic approximation and the decoupling between rotational and vibrational motion break down.

A Calculation of C_p and Comparison to Experiment

§2. *A Review of the Equations for C_p.* I know that for an ideal gas

$$C_p = C_v + R \tag{10.1}$$

the heat capacity at constant volume is

$$C_v = (C_v)_{\mathrm{t}} + (C_v)_{\mathrm{v}} + (C_v)_{\mathrm{r}} \tag{10.2}$$

The translational contribution $(C_v)_t$ is (see Eq. 9.25)

$$(C_v)_t = \frac{3R}{2} \tag{10.3}$$

The vibrational contribution $(C_v)_v$ is (see Eq. 9.41)

$$(C_v)_v = R\left(\frac{T_v}{T}\right)^2 \frac{\exp[T_v/T]}{(\exp[T_v/T]-1)^2} \tag{10.4}$$

where T_v is the vibrational temperature (values are given in Table 9.1). The rotational contribution $(C_v)_r$ is (see Eq. 9.58)

$$(C_v)_r = R \qquad \text{if } T \gg T_r \tag{10.5}$$

with T_r the rotational temperature. The condition $T \gg T_r$ is easily fulfilled since the largest value of T_r is 87.547 K (see Table 9.1).

The energy needed to excite the electrons in H_2 is very high, so I can neglect the electronic contribution to C_v. There is never a nuclear contribution to C_v except at very low temperatures when the nuclear spin is coupled with the rotational motion (see Supplement 9.1).

Note that $(C_v)_t$ and $(C_v)_r$ *are the same for all diatomic molecules.* Therefore, differences between the values of C_p for various diatomics must come from $(C_v)_v$.

Combining Eqs 10.1–10.5 gives

$$\frac{C_p}{R} = \frac{7}{2} + \left(\frac{T_v}{T}\right)^2 \frac{\exp[T_v/T]}{(\exp[T_v/T]-1)^2} \tag{10.6}$$

If I use $R = 8.1345$ J/mol K, to get C_p in J/mol K, I obtain

$$C_p = 29.1045 + 8.31451\left(\frac{T_v}{T}\right)^2 \frac{\exp[T_v/T]}{(\exp[T_v/T]-1)^2} \text{ J/mol K} \tag{10.7}$$

§3. *Qualitative Features of the Magnitude of C_p.* Table 10.1 shows that the values of C_p for HCl, H_2, O_2, and N_2 at 298 K are all very close to 29 J/mol K (NO, CO, HI also belong to this group). The values of C_p for Cl_2, I_2, and K_2 are considerably larger. Why this difference? What makes a diatomic molecule belong to one group or the other?

A look at Eq. 10.7 gives us a hint: C_p is equal to 29 J/mol K if the last term in that equation is negligible. If we examine Table 10.1 we notice that those molecules for which $C_p \simeq 29$ have large vibrational temperatures T_v. This suggests that these two facts may be connected. Indeed they are. If

$$T \ll T_v \tag{10.8}$$

then $T_v/T \gg 1$ and

$$\exp[T_v/T] - 1 \approx \exp[T_v/T] \tag{10.9}$$

Using this result in Eq. 10.7 leads to

$$R\left(\frac{T_v}{T}\right)^2 \frac{\exp[T_v/T]}{\left(\exp[T_v/T] - 1\right)^2} \approx R\left(\frac{T_v}{T}\right)^2 \exp[-T_v/T] \tag{10.10}$$

When $T_v/T \gg 1$, this vibrational contribution to C_v becomes very small. In that case, $C_p \approx 29.1$ J/mol K (see Eq. 10.7).

As I already mentioned, the vibrational temperatures T_v for H_2, N_2, O_2, HCl, CO, NO, HI are all much larger than 298 K (see Table 10.1) and the C_p of all these molecules is very close to 29.1 J/mol K, as this analysis predicts.

If the vibrational temperature of a molecule does not satisfy the condition $T \ll T_v$, the temperature-dependent term in Eq. 10.7 cannot be neglected. It is easy to see that this term is positive at all temperatures and therefore the heat capacity of such molecules is larger than 29.1 J/mol K. This is why C_p for Br_2, I_2, etc. are substantially larger than 29.1 J/mol K.

Also note that because the last term in Eq. 10.7 is positive, theory predicts that 29.1 J/mol K is the smallest heat capacity at constant pressure that a gas of diatomic molecules can have. If someone measures a smaller value (for an ideal gas) then either the experiment is wrong or statistical mechanics is in big trouble. If you ever encounter this situation, bet on the side of theory.

Exercise 10.1

Calculate the temperature at which C_p of CO differs from 29.1 J/mol K by 1%. Does C_p grow with temperature or become smaller?

§4. *What are the Largest and the Smallest Values of C_p?* Many people are trying to find materials for which some properties have extreme values. For example, one might wish to create an electrically insulating, solid film with a very large dielectric constant or with a very small index of refraction. Such people often ask the theorists: what is the largest dielectric constant a solid can have? Or what is the smallest refractive index a liquid can have?

These are very sensible questions. If we already have a material whose dielectric constant reached the highest possible value, we should stop trying to use chemical modifications to make the dielectric constant higher. If a device requires a film with a dielectric constant higher than the highest possible value, we should cut our losses before we lose all our money by trying to develop it.

I am not aware of any application that would require us to know the smallest or the largest heat capacity a gas of diatomic molecules can have. It is, however, a good academic exercise to see whether the theory developed here is capable of answering this question.

We have already shown in §3 that the smallest value of C_p is 29.1 J/mol K. Next I want to answer the following question: given a fixed temperature, what is the largest value of C_p that an ideal gas of diatomics can have and how can I reach that value? The secret lies within the last term of Eq. 10.7. In this equation T_v is the only parameter that I can change, by changing the molecule in the gas. Therefore, I can answer the question by figuring out how

$$I \equiv 8.31451 \left(\frac{T_v}{T}\right)^2 \frac{\exp[T_v/T]}{(\exp[T_v/T]-1)^2} \text{ J/mol K}$$

changes with T_v. The easiest way of determining how the function

$$f(x) \equiv x^2 \frac{\exp[x]}{(\exp[x]-1)^2}$$

behaves is to plot it. In Workbook SM10.1 I made such a plot and found that the function is zero for large values of x (it is very small by the time $x = 10$), increases as x decreases, and has a maximum value of one when $x = 0$.

Exercise 10.2

Use calculus to prove the statements made above about the function $f(x)$.

This means that the largest value C_p can possibly have is reached when $x \equiv T_v/T$ is zero, in which case (see Eq. 10.7) $C_p \to 29.1{+}8.314 = 37.414$ J/mol K. What is this telling me? First, there is an upper limit on the values of C_p a gas of diatomics can have. At any given temperature, this limit is reached by a diatomic for which $T_v = 0$ (this makes $x \equiv T_v/T = 0$). Since T_v is proportional to the vibrational frequency, it is physically impossible to find a diatomic whose vibrational frequency is zero! The best we can do is to work with a diatomic whose T_v is very small. With guidance from theory, the quest of finding the gas with the highest value of C_p has become the quest of finding the diatomic with the smallest value of T_v.

Working with diatomics whose T_v is small has limitations. As a rule, when the vibrational frequency (i.e. T_v) is very small, the bond is weak and the diatomic dissociates easily. The smaller T_v is, the lower the temperature at which the dissociation of the diatomic is substantial. If the molecule dissociates, it produces atoms and these have a low C_p: dissociation defeats the purpose of having high C_p! A gas with low T_v has a high C_p, but only if the temperature of the gas is low. Nature puts all kinds of obstacles in our way, to test our ingenuity.

The data in Table 10.1, which gives C_p at $T = 298$ K, are bitter-sweet. Sweet because they show that the C_p of the gases with small T_v is much larger than that of the gases having large T_v. Bitter because the C_p of I_2 is 38.081 J/mol K and exceeds the maximum value predicted by theory! Moreover T_v for K_2 is smaller than that of I_2; according to our theory, K_2 should have a higher C_p than I_2 , but it does not!

As T_v becomes smaller the predictions of the theory become a bit shakier. This happens for two reasons: the bond of a molecule having a small T_v is softer, and the harmonic approximation and the assumption that rotations are decoupled from vibrations start to break down. Moreover, these molecules tend to have low-lying electronic states (I_2 is colored), a feature that we have ignored in our calculations. Therefore, in the limit when T_v is small, the theory we have used for finding the maximum value of C_p starts to make errors. Note, however, that the errors are very small; not more than 3 or 4%. There are few things in life that can be predicted with this accuracy.

Exercise 10.3

(Difficult but interesting) Find the energies of the excited states for I_2 and Br_2 in the NIST database or Huber and Herzberg (see page 126) or Herzberg (see page 131). Derive a formula for C_p that takes into account the electronic degrees of freedom. Follow the procedure used for a gas of atoms. A subtle modification is needed because the vibrational frequency and the rotational constants of an electronically excited molecule differ from those of the ground state. You can find these quantities in Huber and Herzberg. Do a numerical calculation of C_p to see whether including the electronic excitation explains the deviation from the rule regarding the largest value of C_p.

---

§5. *Compare the C_p of N_2 and K_2 to Experiment.* So far I have used Eq. 10.7 to look for trends and answer qualitative questions. Next I use it to calculate C_p and compare the results with the values given in the NIST database. I will do this for N_2 and K_2. These molecules represent two extremes. The N_2 molecule is one of the stiffest diatomics; it has a vibrational frequency of 2359.61 cm^{-1}, which corresponds to a vibrational temperature of 3392.01 K (see Table 10.2). Because of this, I expect that the harmonic approximation and the decoupling of rotation from vibration work well, except for fairly high energies. Since the high energies contribute to the partition function only at high temperatures, I expect to see deviations from the exact results at high temperatures only.

The lowest energy needed for exciting the electrons in the nitrogen molecule is 69,290 cm^{-1}. This corresponds to a huge temperature (please calculate how large). Therefore, I don't need to worry about contributions from electronic excitations. The nuclear spin degrees of freedom contribute to the C_p of N_2 only at very low temperatures. Therefore, I can ignore them.

The vibrational frequency of K_2 is very low (92.021 cm^{-1}) and the vibrational temperature is 132.38 K. The bond in K_2 is very soft, and anharmonic effects will

Table 10.2 The vibrational contribution to C_v (second term in the right-hand side of Eq. 10.7) for several diatomic molecules at $T = 298$ K.

Molecule	T_v (K)	$(C_v)_v$ (J/mol K)
CO	3121.48	2.58×10^{-2}
N_2	3392.01	1.23×10^{-2}
HCl	4301.38	9.33×10^{-4}
H_2	6338.20	2.18×10^{-6}
Cl_2	807.30	4.66
I_2	308.65	7.61
K_2	132.38	8.18

Table 10.3 A comparison of calculated and measured C_p for N_2. Calculations are done in Workbook SM10.6 ; experimental data are from Din's book.

Temperature, T (K)	C_p (J/mol K)		Percentage error
	Experimental	Calculated	
200	28.97	29.10	−0.46
250	28.95	29.10	−0.53
300	28.98	29.12	−0.47
350	29.03	29.15	−0.41
400	29.09	29.23	−0.47
450	29.15	29.35	−0.70
500	29.21	29.54	−1.10
550	29.27	29.77	−1.70
600	29.33	30.04	−2.40
650	29.39	30.34	−3.10
700	29.45	30.66	−3.90

be important at much lower energies than for N_2. The same can be said about the errors caused by decoupling rotation from vibration. I suspect that, at high temperature, the calculated C_p for K_2 will differ from the observed value.

The C_p of N_2 is calculated and compared to experimental results in Workbook SM10.6. In making such comparisons, one should examine the data critically. The discrepancy between the calculations and the results of measurement can be due to experimental errors. To show you that this can happen, I used two sources of data. One is the book by Din[a]. The other is the JANAF data, reproduced at the NIST website. The calculated and measured values of C_p are given in Tables 10.3 and 10.4.

The two tables display a comedy of errors. The calculations agree with Din's data up to temperatures of 450 K and deteriorate as the temperature increases. The comparison with the JANAF data gives the largest errors at 500 K and 600 K and the calculated values agree well with those in the database at higher temperatures. The two data sets do not agree well with each other. I strongly suspect that the JANAF data is not measured but it is calculated from statistical mechanics. The difference

[a] *Thermodynamic Functions of Gases, Vol.* 3, F. Din, editor, Butterworth, London, 1961.

Table 10.4 A comparison of calculated and measured C_p for N_2. Data from JANAF tables on the NIST website.

Temperature, *T* (K)	C_p (J/mol K)		Percentage error
	Experimental	Calculated	
300	28.88	29.12	−0.820
400	29.35	29.23	0.420
500	29.91	29.54	1.200
600	30.47	30.04	1.400
700	31.02	30.66	1.200
800	31.55	31.32	0.730
900	32.06	31.96	0.320
1000	32.54	32.55	−0.027
1100	32.99	33.08	−0.270
1200	33.42	33.55	−0.380
1300	33.81	33.96	−0.430
1400	34.18	34.31	−0.390
1500	34.53	34.62	−0.270
2000	35.91	35.68	0.640

between our calculations and theirs may come from using slightly different atomic data and perhaps they use a formula for rotational and vibrational energies that includes rotation–vibration coupling and anharmonicity. Din's data were measured in the laboratory. I suspect that the discrepancy between theory and experiment grows with temperature because of the breakdown in the harmonic approximation and of the decoupling of rotation from vibration.

§6. *Results for* K_2. The results for K_2 are shown in Table 10.5 from Workbook SM10.7. The errors at high temperature are substantial. The excitation energy of the lowest electronic state is 11,682 cm^{-1}. It is unlikely that the neglect of the electronic excitations is the source of error. We must blame the harmonic approximation and the decoupling of vibrations from rotations.

§7. *The Magnitude of* C_p*: Summary.* In testing how well the theory predicts the values of C_p, you have learned several things. Except for stiff molecules, the theory does not work well at high temperatures. It is likely that this is caused by a

Table 10.5 A comparison of calculated and measured C_p for K_2. Data from the JANAF tables.

Temperature, T (K)	C_p (J/mol K)		Percentage error
	Experimental	Calculated	
300	37.97	37.28	−1.80
400	38.28	37.34	−2.40
500	38.75	37.37	−3.60
600	38.94	37.38	−4.00
700	38.75	37.39	−3.50
800	38.20	37.40	−2.10
900	37.34	37.40	0.17
1000	36.25	37.41	3.20
1100	35.01	37.41	6.80
1200	33.72	37.41	10.90
1300	32.47	37.41	15.20

breakdown of the harmonic approximation and of the decoupling of rotations from vibrations.

When making such comparisons to experiment, you should examine the data critically. Have the authors determined error bars? Is the method of measurement accurate? Even for N_2, a simple molecule that is easy to work with, the data differ from source to source. It must be considerably harder to get good data for K_2, which dissociates easily and tends to form clusters (K_4, K_6, etc.).

Exercise 10.4

If you think that the discrepancy between the calculated C_p for K_2 and the measured values is due to errors in experiment, try I_2. This is also a soft molecule. Compare your results with the NIST data. You will find quite a bit of disagreement with experiment (see Workbook SM10.8). The experimental difficulties mentioned for K_2 are not present for I_2.

To remove computational error, we may get from quantum mechanics better formulae for the rotational and vibrational energies. Such formulae exist and you may try using the one suggested in the next exercise.

Exercise 10.5

Spectroscopists have developed a more accurate formula for the vibrational–rotational energy of K_2. It is (from Huber and Herzberg)

$$\begin{aligned}\varepsilon(v,j) &= \omega_e(v+\tfrac{1}{2}) - \omega_e x_e(v+\tfrac{1}{2})^2 + \omega_e y_e(v+\tfrac{1}{2})^3 + \omega_e z_e(v+\tfrac{1}{2})^4 \\ &\quad + \left[B_e - \alpha_e(v+\tfrac{1}{2}) + \gamma_e(v+\tfrac{1}{2})^2\right] j(j+1) \\ &\quad - \left[D_e + \beta_e(v+\tfrac{1}{2})\right] j^2(j+1)^2 + H_v j^3(j+1)^3 \end{aligned} \qquad (10.11)$$

The values of the constants are

$$\begin{aligned}\omega_e &= 92.021\ \text{cm}^{-1} \\ \omega_e x_e &= 0.2829\ \text{cm}^{-1} \\ \omega_e y_e &= -2.055 \times 10^{-3}\ \text{cm}^{-1} \\ B_e &= 0.056743\ \text{cm}^{-1} \\ \alpha_e &= 0.000165\ \text{cm}^{-1} \\ \gamma_e &= -7.2 \times 10^{-6}\ \text{cm}^{-1} \\ D_e &= 8.63 \times 10^{-8}\ \text{cm}^{-1}\end{aligned}$$

(a) Use Eq. 10.11 to calculate the vibrational–rotational partition function

$$q_{\text{rv}} = \sum_{j=0}^{\infty}\sum_{v=0}^{\infty}(2j+1)\exp\left[-\frac{\varepsilon(v,j)-\varepsilon(0,0)}{k_B T}\right] \qquad (10.12)$$

Compare the results to those obtained by using the partition functions q_{v} and q_{r} given in Chapter 9 (they were obtained by making the harmonic approximation and assuming a decoupling of vibration from rotation).

(b) Calculate C_p by using the partition function given in part (a). To do this, use

$$C_p = \frac{5R}{2} + \frac{\partial u_{\text{vr}}}{\partial T}$$

and get u_{vr} from

$$u_{\text{vr}} = RT^2 \frac{\partial}{\partial T} \ln q_{\text{vr}}$$

Write a program that performs the derivatives and calculate C_p by using Eq. 10.12 for q_{vr}. When you evaluate C_p make sure that you only use in the sum in Eq. 10.12 values of $\varepsilon(v,j)$ that are smaller than the dissociation energy, which is $D_0 = 0.514$ eV.

(c) Eq. 10.11 is reliable only for v up to about 3 and j between 0 and 20. Analyze whether in your calculation, values of v and j outside this range contribute to the partition function. If they do, the calculation is not very reliable.

(d) Compare the results obtained in part (b) to the data given in Table 10.5.

A Comparison of the Calculated Entropy of O_2 to Experiment

§8. *Introduction.* I will further test the theory by calculating the entropy of $^{16}O_2$ and comparing the results to the experimental values. Here is a summary of the equations used. (I have taken them from Chapter 9.)

$$s = s_t + s_v + s_r + s_n + s_e \tag{10.13}$$

$$\frac{s_t}{R} = \ln\left(\frac{RT}{pN_A\Lambda^3}\right) + \frac{5}{2} \tag{10.14}$$

$$\Lambda = \sqrt{\frac{h^2}{2\pi m k_B T}} \tag{10.15}$$

$$\frac{s_v}{R} = \frac{T_v}{T}\frac{1}{\exp[T_v/T] - 1} \tag{10.16}$$

$$\frac{s_r}{R} = \ln\left(\frac{T}{\sigma T_r}\right) \tag{10.17}$$

$$\frac{s_n}{R} = \ln\left([2I+1]^2\right) = 0 \tag{10.18}$$

$$\frac{s_e}{R} = \ln 3 \tag{10.19}$$

The notation has been introduced in Chapter 9 and I don't review it here. The nuclear contribution to entropy, s_n, is zero because the spin of the ^{16}O nucleus is $I = 0$. Oxygen is an unusual molecule: it is a triplet in the ground state and the degeneracy of the ground electronic state is 3. Since the diatomic is homonuclear, $\sigma = 2$.

Table 10.6 Various contributions to entropy for O_2. s_t is translational, s_v is vibrational, s_r is rotational, s_n is nuclear, s_e is electronic, and s is the total entropy. These values were calculated in Workbook SM10.8.

Temperature, T(K)	s_t (J/mol K)	s_v (J/mol K)	s_r (J/mol K)	s_n (J/mol K)	s_e (J/mol K)	s (J/mol K)
300	152.208	0.0364789	43.8902	0	9.13442	205.269
500	162.826	0.4934780	48.1375	0	9.13442	220.592
700	169.820	1.4210400	50.9351	0	9.13442	231.311
900	175.044	2.5183100	53.0246	0	9.13442	239.722
1100	179.215	3.6156300	54.6931	0	9.13442	246.659
1300	182.688	4.6512800	56.0821	0	9.13442	252.556
1500	185.662	5.6092200	57.2719	0	9.13442	257.678
1700	188.264	6.4902600	58.3126	0	9.13442	262.201
1900	190.576	7.3008000	59.2373	0	9.13442	266.249

§9. *Units.* Since Eqs 10.14 and 10.16–10.19 give various contributions to the entropy divided by the gas constant R, I can get entropy in units of J/mol K by taking $R = 8.134$ J/mol K. Note, however, that R also appears in the argument of the logarithm in the expression for s_t. Under the logarithm, the pressure is 1 bar, and Λ^3 has units of cm^3 (using CGS units); therefore, I take there $R = 83.145$ bar cm^3/mol K. This complication aside, the calculation of the entropy is straightforward and it is performed in Workbook SM10.8.

The various contributions to entropy are given in Table 10.6.

§10. *Can we Understand These Results?* Table 10.6 shows that the largest contribution to entropy comes from translation and the next largest from rotations. The vibrational contribution is very small, even smaller than that caused by the electronic degeneracy of the ground state.

In Chapter 3 we connected entropy with information, which in turn reflected our ability to guess the state of the system. Will this connection help us understand these results? The word "understand" is a tricky one: we use it without giving much thought to its meaning. What does it mean *to understand* why a certain contribution to entropy is smaller than another contribution? Dirac has given a meaning to this word when he said that understanding a quantity means that you can predict its behavior without having to do elaborate calculations. You may disagree with this definition of understanding, but I am going to use it here.

In Chapter 3 I have shown that entropy is connected to my ability to guess the state of a molecule in the system. Let us denote the possible energies a molecule can take by ε_0, ε_1, ε_2, ... These values are ordered: ε_0 is the smallest, ε_1 is either equal to ε_0 or larger, etc.

You remember that some of the states of a molecule can have the same energy (they are degenerate states). For example, there are $2j + 1$ rotational states having the energy $j(j + 1)B$. The vibrational states are not degenerate. The spin states of the electrons in the ground state of the O_2 molecule are triply degenerate (there are three spin states having the same energy).

The probability that a molecule in the gas is in the state i is (see Chapter 1)

$$p_i = \frac{\exp\left[-\varepsilon_i/k_B T\right]}{\sum_{j=0}^{\infty} \exp\left[-\varepsilon_j/k_B T\right]} \tag{10.20}$$

The sum in Eq. 10.20 is over all the states of the molecule. You can easily see that Eq. 10.20 can be written as

$$p_i = \frac{\exp\left[-\theta_i/T\right]}{1 + \sum_{j=0}^{\infty} \exp\left[-\theta_j/T\right]} \tag{10.21}$$

where

$$\theta_i \equiv \frac{\varepsilon_i - \varepsilon_0}{k_B} \tag{10.22}$$

θ_i is called the *excitation temperature* of the state i. Note that $\theta_0 = 0$ and that degenerate states have the same excitation temperature.

The entropy of 1 mole of gas is given by (see Chapter 3)

$$s = -R \sum_{i=0}^{\infty} p_i \ln p_i \tag{10.23}$$

where the sum is again over all states.

Let us use this framework to examine the vibrational contributions to entropy. The energies are (see Chapter 9)

$$\varepsilon_\alpha = \hbar\omega \left(\alpha + \frac{1}{2}\right) - D_0 \qquad \alpha = 0, 1, \ldots \tag{10.24}$$

The excitation temperatures are

$$\theta_\alpha = \frac{\varepsilon_\alpha - \varepsilon_0}{k_B} = \alpha T_v \tag{10.25}$$

where $T_v = \hbar\omega/k_B$. If $T_v \gg T$ then $\theta_\alpha \gg T$ for $\alpha = 1, 2, \ldots$ and (see Eq. 10.21) $p_0 \approx 1, p_1 \approx 0, p_2 \approx 0, \ldots$. Using this in Eq. 10.23 for entropy gives $s \approx 0$ when $T_v \gg T$.

This result makes sense. When $\theta_\alpha \gg T$, $\alpha = 1, 2, \ldots$, we can guess with a high degree of certainty that any molecule in the gas is in the ground state $\alpha = 0$, so the entropy is very small (no "disorder" in our information).

As the temperature increases and starts being larger than some excitation temperature, the entropy increases. For example, if $\theta_3 \leq T \ll \theta_4$, then p_0, p_1, p_2, p_3 have sizeable values, and the rest are close to zero. Our ability to guess the state of a molecule in the gas decreases and the entropy increases. Eq. 10.23 shows that this is the case.

For oxygen, T_v is 2273.64 K, which makes $\theta_1 = 2273.64$ K (see Eq. 10.25). At normal temperatures, $T < \theta_1$ and this makes the vibrational entropy nearly zero. For soft molecules, such as I_2 and K_2, the vibrational temperature is small and θ_1 and even θ_2 are within the temperature range normally used in the laboratory. The vibrational entropy of these molecules is not zero.

The same analysis can be applied to rotational degrees of freedom. Here

$$\theta_j = \frac{\varepsilon_j - \varepsilon_0}{k_B} = \frac{j(j+1)B}{k_B} = j(j+1)T_r \tag{10.26}$$

Since the highest rotational temperature T_r (for H_2) is 87.55 K and it is much smaller for all other molecules, the condition $\theta_j \ll T$ is satisfied by a large number of states. For all these states p_j is sizeable and therefore the entropy is rather high. The same kind of argument will lead us to conclude that the translational contribution to entropy is even higher than the rotational contribution since more translational states satisfy $\theta_j \ll T$.

Finally, the electronic state is triply degenerate, which means that the three spin states are equally probably, $p_i = \frac{1}{3}$, and the electronic entropy is

$$-R\sum_{i=1}^{3} p_i \ln p_i = -R \times 3 \ln \frac{1}{3} = 3R \ln 3$$

Table 10.7 A comparison of the calculated entropy of O_2 with the data given at the NIST website.

Temperature, *T* (K)	Entropy, *s* (J/mol K)		Percentage error
	Experimental	Calculated	
300	205.2	205.269	0.033
500	220.8	220.592	−0.094
700	231.6	231.311	−0.125
900	240.0	239.722	−0.116
1100	246.9	246.659	−0.098
1300	252.8	252.556	−0.097
1500	258.0	257.678	−0.125
1700	262.6	262.201	−0.152
1900	266.8	266.249	−0.207

As you can see, it is possible to say quite a bit about the magnitude of different contributions to entropy without performing detailed calculations. Dirac would say that statistical mechanics helps us understand entropy.

The results of the calculations are compared to data in Table 10.7.

§11. *Summary.* The theory is qualitatively correct, and it is quantitatively accurate at temperatures that do not exceed by much the vibrational temperature. The errors, at high temperatures, come from the harmonic approximation, the decoupling of rotation from vibration, and in a few cases from the neglect of the electronic excitation of the molecule.

Problems

Problem 10.1

(a) Calculate the entropy of H_2. Get T_v and T_r from Table 9.1. The nuclear spin of H is $\frac{1}{2}$ and the degeneracy of the electronic ground state is $g_0^e = 1$. (See results in Workbook SM10.9.) Examine all individual contributions (from translation, vibration, etc.). (b) Explain the results. (c) Compare to the data given in the NIST tables.

Problem 10.2

The potential energy of H_2 is the same as that of D_2 and of HD. The nuclear spin of H is $\frac{1}{2}$ and that of D is 1. (a) Find T_v and T_r for H_2 in Table 9.1. From the data for H_2, calculate T_v and T_r for HD and D_2. (b) Calculate the chemical potential μ of H_2, HD, and D_2. Analyze how each contribution to μ changes as you go from H_2 to HD to D_2. The degeneracy of the electronic ground state of these molecules is $g_0^e = 1$.

Problem 10.3

Compare the chemical potential of H_2 to that of O_2. Try to guess beforehand, for each contribution to μ, whether the result for H_2 is larger or smaller than that for O_2.

Problem 10.4

Analyze where the differences between the entropies of $^{16}O^{16}O$ and $^{16}O^{17}O$ come from — translation, vibration, nuclear, ...? Try to guess, for each contribution, which molecule has the larger entropy.

Problem 10.5

In Table 10.8, I have collected experimental data for the heat capacity C_p, the entropy s, and the change in enthalpy $h(T) - h(298)$ for H_2 and K_2. Use the equations derived in this chapter to calculate these quantities. Compare your results to the experimental values and find out how large the errors of the calculations are.

Problem 10.6

Go to the NIST website webbook.nist.gov/chemistry/. Look for thermodynamic data for Na_2. You will find there data for $X \equiv -(G^0 - H^0_{298})/T$ at $p = 1$ bar. G^0 is the chemical potential μ and H^0_{298} is the enthalpy h at 298 K. Calculate $\mu(T) - \mu(298)$ from these data. Then use statistical mechanics to calculate the same quantity. Use the molecular constants given in Table 9.1.

Problem 10.7

The Morse potential energy of CO is $V(r) = D_0\left(1 - \exp[-a(r - r_0)]\right)^2$, with $D_0 = 11.2254$ eV, $a = 3.25145$ Å^{-1}, and $r_0 = 1.1283$ Å. Use this information to

Table 10.8 Experimental data for heat capacity at constant pressure $C_p(T)$, entropy $s(T)$, and the difference $h(T) - h(298)$ (where h is enthalpy), for H_2 and K_2.

Temperature, T(K)	H_2 $C_p(T)$ (J/mol K)	H_2 $s(T)$ (J/mol K)	H_2 $h(T)-h(298)$ (KJ/mol)	K_2 $C_p(T)$ (J/mol K)	K_2 $s(T)$ (J/mol K)	K_2 $h(T)-h(298)$ (KJ/mol)
298	28.84	130.7	−0.00	37.98	249.7	−0.00
300	28.85	130.9	0.05	37.97	249.9	0.07
400	29.18	139.2	2.96	38.28	260.9	3.88
500	29.26	145.7	5.88	38.75	269.5	7.73
600	29.32	151.1	8.81	38.94	276.6	11.62
700	29.44	155.6	11.75	38.75	282.6	15.51
800	29.63	159.5	14.70	38.20	287.7	19.36
900	29.88	163.1	17.68	37.34	292.2	23.14
1000	30.21	166.2	20.68	36.25	296.0	26.82
1100	30.58	169.1	23.72	35.01	299.4	30.38
1200	30.99	171.8	26.80	33.72	302.4	33.82
1300	31.42	174.3	29.92	32.47	305.1	37.13
1400	31.86	176.6	33.08			
1500	32.30	178.8	36.29			

calculate the chemical potential for temperatures between 298 K and 2000 K. Compare the results with data for $\mu(T) - \mu(298)$ from the NIST database webbook.nist.gov/chemistry/). (In those tables, you will find $-(G^0 - H^0_{298})/T$ where G^0 is μ; calculate $\mu(T) - \mu(298)$ from this.)

11

CHEMICAL EQUILIBRIUM

Introduction

§1. *The Problem.* If you perform the reaction

$$A + B \rightleftharpoons C + D$$

in a closed container, at constant temperature and pressure, the number of moles of A will decrease and then reach a constant value. When the composition of the system stops changing in time the system reached *chemical equilibrium*. One of the important tasks of physical chemistry is to calculate how the concentration of the reactants and products in the equilibrated system depend on pressure, temperature, and initial concentrations.

You have already learned, when you studied thermodynamics, how to use thermodynamic data to perform such calculations. The central quantity in that theory was the *equilibrium constant* of the reaction.

In this chapter, you will learn how to use statistical mechanics to express the equilibrium constant in terms of the molecular properties of the participants in the reaction: the masses, the vibrational frequency, the rotational constant, and the binding energy.

We study here only reactions that involve atoms and diatomic molecules, because you have not learned how to deal with larger molecules. The method is, however, general and this kind of calculation can be performed for gas-phase reactions involving polyatomic molecules.

The extension to reactions in condensed phases (where the molecules interact with each other) is straightforward, but its implementation requires using large computer programs. Right now the most serious impediment to calculating equilibrium constants in condensed phases is a lack of accurate knowledge of the interaction energies between the atoms of the same molecule and between the various molecules in the system. Computational chemistry is making rapid progress in this direction.

A Crude Model of Chemical Equilibrium

§2. *A Brief Summary.* I will begin with a crude model of a chemical reaction which assumes that the system has two states: one represents the reactants, the other the products. In this model, the reactants and the products differ only through their ground-state energies; the fact that the molecules vibrate, translate, and rotate is ignored.

In spite of its crudity this model captures an essential fact about the equilibrium constant: its magnitude is dominated by the binding energy of the molecule. The vibrational, rotational, and translational energies do matter, but their inclusion (see §12) amounts to a relatively small correction to the crude theory.

§3. *The Simplified Model.* I start by assuming that I can capture the essence of the problem by pretending that each molecule in the system has only two important states. In one, which I call d (for "diatomic"), the molecule is bound and has the energy

$$E_d = -D_0 \tag{11.1}$$

Here D_0 is the dissociation energy, and it is a positive number.

In the other state, which I label a (for 'atomic'), the molecule is dissociated and has the energy

$$E_a = 0 \tag{11.2}$$

Of course, the molecule and the atoms move throughout the box and the molecule vibrates and rotates. The energies of these motions are sure to play a role in an accurate theory. I ignore them for now, to obtain the simplest possible theory of dissociation.

§4. *The Equilibrium Constant.* You learned in thermodynamics that the equilibrium constant K of this reaction is:

$$K = \frac{(x_a)^2}{x_d} \tag{11.3}$$

where x_a and x_d are the molar fractions of the atoms and the diatomic. We can calculate the molar fractions if we know the number N_a of atoms, and the number N_d of diatomic molecules in the system, at equilibrium.

Exercise 11.1

Try deriving an expression connecting the equilibrium constant to the energies E_a and E_d, without reading what follows. This is a good test of your understanding of the material you have studied so far.

The probabilities P_a and P_d that a molecule is in state a or in state d, when *the system is at equilibrium*, are

$$P_a = \frac{\exp\left[-\beta E_a\right]}{Q} = \frac{\exp\left[-\beta \cdot 0\right]}{Q} = \frac{1}{Q} \tag{11.4}$$

and

$$P_d = \frac{\exp\left[-\beta E_d\right]}{Q} = \frac{\exp\left[-\beta(-D_0)\right]}{Q} = \frac{\exp\left[\beta D_0\right]}{Q} \tag{11.5}$$

The notation

$$\beta = \frac{1}{k_B T}$$

is very often used in statistical mechanics.

The partition function Q for the two-state model is given by

$$Q = \exp[-\beta E_a] + \exp[-\beta E_d]$$

$$= 1 + \exp[-\beta E_d] = 1 + \exp[\beta D_0] \tag{11.6}$$

This equation is correct only for a model consisting of non-interacting particles that have two states (i.e. dissociated (state a) or bound (state d)).

Using Eq. 11.6 in Eqs 11.4 and 11.5, I find

$$P_a = \frac{1}{1 + \exp[\beta D_0]} \tag{11.7}$$

$$P_d = \frac{\exp[\beta D_0]}{1 + \exp[\beta D_0]} \tag{11.8}$$

I will assume that the system contained initially (i.e. before the reaction took place) N diatomics, which were allowed to dissociate and reach equilibrium. The number of diatomic molecules at equilibrium is:

$$N_d = NP_d = \frac{N \exp[\beta D_0]}{1 + \exp[\beta D_0]} \tag{11.9}$$

The number of molecules that have dissociated is $(1 - P_d)N = P_a N$. Each dissociated molecule creates two atoms, so the number of atoms at equilibrium is

$$N_a = 2P_a N = \frac{2N}{1 + \exp[\beta D_0]} \tag{11.10}$$

The molar fractions at equilibrium are

$$\begin{aligned} x_a &\equiv \frac{n_a}{n_a + n_d} = \frac{N_a/N_A}{(N_a/N_A) + (N_d/N_A)} \\ &= \frac{N_a}{N_a + N_d} = \frac{2}{2 + \exp[\beta D_0]} \end{aligned} \tag{11.11}$$

The number of moles of atoms and diatomics are

$$n_a = \frac{N_a}{N_A}$$

and

$$n_d = \frac{N_d}{N_A}$$

where N_A is Avogadro's number. These expressions are used to obtain the second equality in Eq. 11.11; Eqs 11.9 and 11.10 lead to the last equality. Similarly,

$$x_d = \frac{\exp[\beta D_0]}{2 + \exp[\beta D_0]} \tag{11.12}$$

Exercise 11.2

Use Eqs 11.10 and 11.9 to calculate the molar fractions of the diatomic and of the atoms at equilibrium and the equilibrium constant for the reaction $I_2 \rightleftharpoons 2I$. The value of D_0 is given in Table 11.1.

Substituting the molar fractions, given by Eqs 11.11 and 11.12, in Eq. 11.3, I obtain

$$K = \frac{4 \exp[-\beta D_0]}{2 + \exp[\beta D_0]} \tag{11.13}$$

Exercise 11.3

Use the same kind of model to derive the expression of the equilibrium constant for the reaction $A + B \rightleftharpoons AB$.

Table 11.1 The dissociation energy (in electron volts and kelvin) for several diatomic molecules.

Molecule	D_0 (eV)	D_0/k_B (K)
H_2	4.470	51,856
D_2	2.470	28,654
O_2	5.115	59,338
N_2	9.760	113,225
CO	11.090	128,654
I_2	1.540	17,864
K_2	0.514	5,964

Exercise 11.4

Use the same kind of model to derive the expression of the equilibrium constant for the reaction $AB + CD \rightleftharpoons AC + BD$.

§5. *Interpretation.* In this simple model, the equilibrium constant K and the equilibrium concentrations x_a and x_d depend on only one molecular parameter, the dissociation energy D_0. The quantity

$$T_0 = \frac{D_0}{k_B} \tag{11.14}$$

has units of temperature. Since its definition involves the dissociation energy, I will call it the *dissociation temperature*. If

$$T \ll T_0 \tag{11.15}$$

then $\beta D_0 = D_0/k_B T = T_0/T \gg 1$, and $\exp[\beta D_0]$ is much larger than 1. This means that $1 + \exp[\beta D_0] \approx \exp[\beta D_0]$ and (from Eq. 11.8)

$$p_d \approx \frac{\exp[\beta D_0]}{\exp[\beta D_0]} = 1$$

I find that if $T \ll T_0$, practically all the molecules in the system are diatomics. There can be a substantial number of atoms in the gas only if

$$\beta D_0 = \frac{D_0}{k_B T} = \frac{T_0}{T} \leq 1$$

Exercise 11.5

Find out what happens to K, x_a, and x_d when $T \ll T_0$ and when $T \gg T_0$.

Let us take a look at some values of the dissociation energies. Table 11.1 gives the dissociation energies and dissociation temperatures of a few molecules. Dissociation is very effective only if $D_0/k_BT = T_0/T$ is of order 1 or smaller. The table tells me that very high temperatures are needed to satisfy this condition and dissociate a diatomic molecule.

This conclusion is in agreement with our experience. Air consists mainly of O_2 and N_2 and these are very stable. If they were not, life on Earth would be very different. The atoms are very aggressive, chemically, and if there were a lot of them, the world would be full of oxides and nitrites, bacteria (the good ones and the bad ones) would be unable to live, etc. It is, in fact, very difficult to prepare I, H, N, and O atoms; we have to use violent methods, such as electric discharges, very hot metal filaments, UV radiation, electron bombardment, etc.

What this analysis is telling us is very simple. The state in which the two atoms are separated has substantially more energy than the one in which they are bound as a diatomic. Because of this, it is improbable that atoms will be observed, in a gas of diatomics at equilibrium, unless the temperature is very high. The dissociation equilibrium is mainly controlled by the dissociation energy.

There is another way of looking at this, which leads to the same conclusion. The energy needed to dissociate a diatomic must come from collisions with other molecules. If dissociation requires a very large energy (D_0), then the collision partner must be travelling very fast. Finding a molecule with very large kinetic energy is improbable, unless the temperature is very high (remember that the mean translational energy is $3k_BT/2$, which is rather small).

I am sure that you are aware that this model is very crude. It ignores the translational and the rotational energy and the other vibrational states. For more precise calculations we need to include these effects. I will do that next.

A Statistical Mechanical Theory

§6. *Outline.* To construct a statistical mechanical theory of chemical equilibrium, we start with the equilibrium condition derived in thermodynamics, which leads to an equation involving the chemical potentials of the reaction participants. Then, I will replace the chemical potentials in the equilibrium condition with their expressions provided by statistical mechanics. This will express the equilibrium condition in terms of the molecular properties (mass, vibrational frequency, rotational constant, etc.) of the reaction participants.

§7. *A Review of Thermodynamics.* The equilibrium condition for the reaction $A_2 \rightleftharpoons 2A$ is

$$2\mu(A) - \mu(A_2) = 0 \tag{11.16}$$

Here $\mu(A_2)$ is the chemical potential of the diatomic and $\mu(A)$ is that of the atom. Similar conditions apply to other reactions. For example, for an isomerization reaction

$$C \rightleftharpoons D$$

the condition is

$$\mu(C) - \mu(D) = 0;$$

for the bimolecular reaction

$$H_2 + I_2 \rightleftharpoons 2HI$$

the equilibrium condition is

$$2\mu(HI) - \mu(H_2) - \mu(I_2) = 0.$$

For an ideal solution, which is the only one considered here, the chemical potential $\mu(A) \equiv \mu(A; T, p, x(A))$ depends on temperature, pressure, and the molar fraction $x(A)$ of the compound A. As a result, the equilibrium condition $2\mu(A; T, p, x(A)) - \mu(A_2; T, p, x(A_2)) = 0$ connects the equilibrium concentrations $x(A)$ and $x(A_2)$ to the pressure p and the temperature T in the system. To use this equation I need to know how μ depends on these quantities. It has been established empirically that, for an ideal solution,

$$\mu(A_2) = \mu^0(A_2; T, p) + RT\ln(x(A_2)) \tag{11.17}$$

Here, $x(A_2)$ is the molar fraction of the molecules A_2 and $\mu^0(A_2; T, p)$ is the chemical potential of a *pure gas* of diatomic molecules at a pressure p and a temperature T. A similar equation can be written for an ideal gas of atoms:

$$\mu(A) = \mu^0(A; T, p) + RT\ln(x(A)) \tag{11.18}$$

If I use Eqs 11.17 and 11.18 in Eq. 11.16, I obtain (use $\ln(a) - \ln(b) = \ln(a/b)$ and $2\ln(a) = \ln(a^2)$)

$$-RT \ln\left(\frac{x(\mathrm{A})^2}{x(\mathrm{A}_2)}\right) = 2\mu^0(\mathrm{A}; T, p) - \mu^0(\mathrm{A}_2; T, p) \tag{11.19}$$

The quantity

$$\frac{x(\mathrm{A})^2}{x(\mathrm{A}_2)} \equiv K(T, p) \tag{11.20}$$

is the equilibrium constant. By using the notation of Eq. 11.20, I can write Eq. 11.19 as

$$-RT \ln K(T, p) = 2\mu^0(\mathrm{A}; T, p) - \mu^0(\mathrm{A}_2; T, p) \tag{11.21}$$

The right-hand side of this equation is independent of the concentration. Therefore, the left-hand side must also be independent of concentration, even though the expression contains the molar fractions. This means that when a reaction reaches equilibrium, for a given temperature and pressure, the ratio $x(\mathrm{A})^2/x(\mathrm{A}_2)$ is the same no matter what the initial concentrations were when we started the reaction.

When you studied thermodynamics, you used this equation to calculate how the equilibrium concentrations depend on T, p, and the initial concentrations.

§8. *A Statistical Mechanical Theory.* Statistical expressions for the chemical potentials, in terms of molecular properties, were provided in Chapter 9 (for diatomics) and Chapter 6 (for atoms). Before using them, I have to rewrite them and make them look like Eqs 11.17 and 11.18. This rewriting determines expressions for $\mu^0(\mathrm{A}; T, p)$ and for $\mu^0(\mathrm{A}_2; T, p)$. These are used in Eq. 11.21 to give a formula that expresses the equilibrium constant K in terms of molecular parameters.

§9. *The Chemical Potential of the Gas of Atoms.* The chemical potential of 1 mole of gas of atoms A is (see Eq. 6.68)

$$\mu(\mathrm{A}; T, p(\mathrm{A})) = -RT \ln\left(\frac{V}{N(\mathrm{A})\Lambda(\mathrm{A})^3}\right) - RTq_e(\mathrm{A}) - RT \ln(2I_\mathrm{A} + 1) \tag{11.22}$$

where $N(\mathrm{A})$ is the number of A atoms and $\Lambda(\mathrm{A})$ is their de Broglie wavelength, $q_e(\mathrm{A})$ is the electronic partition function (see Eq. 5.34), and I_A is the nuclear spin of the atom A. From the ideal gas equation for a mixture, I have

$$\frac{V}{N(A)} = \frac{k_B T}{p(A)} = \frac{RT}{N_A p(A)}$$

where $p(A)$ is the partial pressure of A and N_A is Avogadro's number. To obtain the last equality I used $N_A k_B = R$. The partial pressure is connected to the molar fraction through $p(A) = x(A)p$. Inserting these expressions in Eq.11.22 leads to

$$\mu(A; T, p) = RT \ln(x(A)) - RT \ln\left(\frac{RTq_e(A)\,(2I_A + 1)}{\Lambda(A)^3 N_A p}\right) \tag{11.23}$$

Comparing Eq. 11.23 to Eq. 11.18, I obtain

$$\begin{aligned} \mu^0(A; T, p) &= -RT \ln\left(\frac{RTq_e(A)\,(2I_A + 1)}{\Lambda(A)^3 N_A p}\right) \\ &\equiv -RT \ln\left[Z(A) q_e(A)\,(2I_A + 1)\right] \end{aligned} \tag{11.24}$$

This is the equation for the chemical potential of a pure gas of atoms at pressure p and temperature T. The quantity Z(A) is defined by this equation and it is

$$Z(A) \equiv \frac{RT}{\Lambda(A)^3 N_A p} \tag{11.25}$$

§10. *The Chemical Potential of the Diatomic.* The chemical potential of an ideal gas of diatomic molecules is (see Chapter 9)

$$\begin{aligned} \mu(A_2; T, p(A_2)) &= -RT \ln\left[\left(\frac{V}{N(A_2)\Lambda(A_2)^3}\right) q_v q_r q_n\right] \\ &= -RT \ln\left(\frac{V}{N(A_2)\Lambda(A_2)^3}\right) - RT \ln\left(q_v q_r q_n\right) \end{aligned} \tag{11.26}$$

I have neglected here the electronic contribution to the chemical potential of the diatomic. q_v, q_r, and q_n are the vibrational, rotational, and nuclear partition functions. The nuclear partition function is $q_n = (2I_A + 1)^2$.

The first term in Eq. 11.26 is similar to the chemical potential of a gas of atoms, given in Eq. 11.22. Performing on this term the same transformations I performed on Eq. 11.22 to obtain Eq. 11.24, *leads to*

$$-RT\ln\left(\frac{V}{N(\mathrm{A}_2)\Lambda(\mathrm{A}_2)^3}\right)=RT\ln(x(\mathrm{A}_2))-RT\ln\left(\frac{RT}{\Lambda(\mathrm{A}_2)^3N_A\,p}\right) \tag{11.27}$$

which is the analog of Eq. 11.24. Using Eq. 11.27 in Eq. 11.26 gives

$$\mu(\mathrm{A}_2;T,p,x(\mathrm{A}_2))=RT\ln\left(x(\mathrm{A}_2)\right)-RT\ln\left(\frac{RT}{\Lambda(\mathrm{A}_2)^3N_A\,p}q_\mathrm{v}q_\mathrm{r}q_\mathrm{n}\right) \tag{11.28}$$

Comparing Eq. 11.28 to Eq. 11.17 shows that

$$\mu^0(\mathrm{A}_2;T,p)=-RT\ln(\mathrm{Z}(\mathrm{A}_2)q_\mathrm{v}q_\mathrm{r}q_\mathrm{n}) \tag{11.29}$$

with

$$\mathrm{Z}(\mathrm{A}_2)\equiv\frac{RT}{\Lambda(\mathrm{A}_2)^3N_A\,p} \tag{11.30}$$

§11. *Calculate the Equilibrium Constant.* I use these expressions for $\mu^0(\mathrm{A}_2;T,p)$ and $\mu^0(\mathrm{A};T,p)$ in Eq. 11.21, which gives the equilibrium constant. I obtain

$$\begin{aligned}-RT\ln K(T,p)&=2\mu^0(\mathrm{A};T,p)-\mu^0(\mathrm{A}_2;T,p)\\&=2\left[-RT\ln\left(\mathrm{Z}(\mathrm{A})q_\mathrm{e}(2I_A+1)\right)\right]-\left[-RT\ln(\mathrm{Z}(\mathrm{A}_2)q_\mathrm{v}q_\mathrm{r}(2I_A+1)^2)\right]\\&=-RT\ln\left(\frac{\mathrm{Z}(\mathrm{A})^2q_\mathrm{e}^2}{\mathrm{Z}(\mathrm{A}_2)q_\mathrm{v}q_\mathrm{r}}\right)\end{aligned} \tag{11.31}$$

Note that the nuclear spin contribution cancels out (I used $q_n=(2I+1)^2$ for the nuclear partition function of the diatomic). This is not surprising: the reaction does not change the degeneracy of the nuclear states.

Eq. 11.31 leads to the main result of this section:

$$K=\frac{\mathrm{Z}(\mathrm{A})^2q_\mathrm{e}^2}{\mathrm{Z}(\mathrm{A}_2)q_\mathrm{v}q_\mathrm{r}} \tag{11.32}$$

This is the equilibrium constant of the reaction $\mathrm{A}_2\rightleftharpoons 2\mathrm{A}$, given in terms of molecular properties only. In the remainder of this chapter, I will use this equation to show how various molecular parameters affect the dissociation equilibrium.

§12. *The Contributions from Rotation, Vibration, and Translation.* For ease of analysis I separate the terms in Eq. 11.32 according to their physical content, by writing it as

$$K = \frac{Z(A)^2 q_e^2}{Z(A_2) q_v q_r} \equiv K_t K_v K_r K_e \tag{11.33}$$

where

$$K_t \equiv \frac{Z(A)^2}{Z(A_2)} \tag{11.34}$$

contains the effect of the translational motion;

$$K_v = \frac{1}{q_v} \tag{11.35}$$

contains the effect of the vibrational motion;

$$K_r = \frac{1}{q_r} \tag{11.36}$$

represents the effect of the rotational motion; and

$$K_e = q_e^2 \tag{11.37}$$

accounts for the electronic excitation of the atoms and the degeneracy of their ground electronic state.

The role of each term is analyzed separately in the following sections.

§13. *Generalization to Other Reactions.* Before proceeding with the analysis we take a side trip, to generalize these results. The procedure for deriving expressions for the equilibrium constant of a reaction is quite general. If I apply it to the reaction

$$H_2 + I_2 \rightleftharpoons 2HI$$

I obtain

$$K_p = \frac{Z(HI)^2 q_v(HI)^2 q_r(HI)^2}{Z(H_2) q_v(H_2) q_r(H_2)\, Z(I_2) q_v(I_2) q_r(I_2)} \tag{11.38}$$

Here Z(HI), q_v(HI), and q_r(HI) are the translational contribution, the vibrational partition function, and the rotational partition function of HI. I ignored the nuclear and the electronic contributions. Note that the contributions from the products are in the numerator and those from the reactants are in the denominator. The contributions from HI are raised to the power 2, because the stoichiometric coefficient of HI in the reaction is 2.

Exercise 11.6

For the reaction $H_2 + I_2 \rightleftharpoons 2HI$, start with the equilibrium condition $2\mu(HI) - \mu(H_2) - \mu(I_2) = 0$ and derive Eq. 11.38 by following a procedure similar to that leading to Eq. 11.32.

How Various Quantities Contribute to the Equilibrium Constant

§14. *The Translational Contribution.* We return now to the dissociation reaction and examine the role of various contributions to the equilibrium constant K (Eq. 11.33). K_t represents contributions due to translational motion. When dissociation takes place, a diatomic molecule is replaced by two atoms. There are more particles translating through the box and their mass is different. K_t tells me how these changes affect the equilibrium constant.

To better understand this point, consider an isomerization reaction $C \rightleftharpoons D$. The reactants and products have the same mass so Z(D) = Z(C). You can verify this by examining the definition of Z. The equilibrium constant is

$$K = \frac{q_v(D)}{q_v(C)} \cdot \frac{q_r(D)}{q_r(C)} \cdot \frac{Z(D)}{Z(C)} = \frac{q_v(D)q_r(D)}{q_v(C)q_r(C)} \tag{11.39}$$

(I have ignored the electronic contribution.) The translational contribution disappears from the equation for the equilibrium constant.

Why does this happen? Should we have expected it? Of course we should. This reaction causes no change in the translational motion in the system: the mass of C is the same as the mass of D, and the mass is the only molecular characteristic in the translational partition function.

Another interesting difference between the two reactions is that the total pressure p does not appear in the expression (Eq. 11.39) for the equilibrium constant of $C \rightleftharpoons D$, but it does appear in the expression (Eq. 11.34) for the translational equilibrium constant of $A_2 \rightleftharpoons 2A$ (to see the effect of pressure, replace Z(A) and Z(A_2)

in Eq. 11.34 with their expressions given by Eqs 11.25 and 11.30). This is consistent with le Chatelier's principle of thermodynamics. A reaction that *increases* the number of molecules will *increase* the pressure in the system. Le Chatelier's principle says that if we increase the pressure, the reaction will evolve in a direction that will frustrate our action: some of the atoms will recombine to produce diatomic molecules, and this will reduce pressure. This means that for this reaction, the equilibrium constant must depend on pressure.

Le Chatelier's principle also tells us that the reaction $C \rightleftharpoons D$ does not cause a pressure change in the system and therefore the equilibrium concentration is not affected by the outside pressure.

§15. *Calculation of* K_t. I will calculate next the values of K_t for the dissociation of H_2 and I_2. I want to see how large the translational contributions are, and how much they differ from one molecule to another.

I will use CGS units in the calculation. The atomic unit of mass is 1.6605×10^{-24} g, $h = 2\pi\hbar$ with $\hbar = 1.0546 \times 10^{-27}$ erg s, and $k_B = 1.3806 \times 10^{-16}$ erg/K. The pressure is normally given in atm and needs to be converted to the CGS unit dyne/cm^2. The conversion factor is 1 atm = 1.0033×10^6 dyne/cm^2. The translational contribution K_t to the equilibrium constant is calculated from Eq. 11.34. I use Eqs 11.25 and 11.30 for Z(A) and Z(A_2) and I calculate Λ from $\Lambda = \sqrt{h^2/2\pi m k_B T}$. The results are shown in Table 11.2 for H_2 and in Table 11.3 for I_2. The calculation was performed in Workbook SM11.2.

The translational contribution to the equilibrium constant is larger for I_2 than for H_2. This reflects the difference in the masses. Indeed, by using the equations

Table 11.2 The translational (K_t), vibrational (K_v), and rotational (K_r) contributions to the equilibrium constant K for the dissociation of H_2.

Temperature, T(K)	K_t	K_v	K_r	K
300	1.43×10^4	5.95×10^{-76}	1.46×10^{-1}	1.24×10^{-72}
700	1.19×10^5	5.76×10^{-33}	6.25×10^{-2}	4.28×10^{-29}
1100	3.68×10^5	3.04×10^{-21}	3.98×10^{-2}	4.44×10^{-17}
1500	7.98×10^5	8.88×10^{-16}	2.92×10^{-2}	2.07×10^{-11}
1900	1.44×10^6	1.28×10^{-12}	2.30×10^{-2}	4.24×10^{-9}
2000	1.86×10^6	1.70×10^{-11}	2.08×10^{-2}	6.58×10^{-7}

Table 11.3 The translational (K_t), vibrational (K_v), and rotational (K_r) contributions to the equilibrium constant K for the dissociation of I_2.

Temperature, T(K)	K_t	K_v	K_r	K
300	2.02×10^{7}	7.91×10^{-27}	8.94×10^{-5}	1.43×10^{-23}
700	1.68×10^{8}	2.80×10^{-12}	3.83×10^{-5}	1.80×10^{-8}
1100	5.19×10^{8}	2.10×10^{-8}	2.43×10^{-5}	2.66×10^{-4}
1500	1.13×10^{9}	1.22×10^{-6}	1.79×10^{-5}	2.47×10^{-2}
1900	2.04×10^{9}	1.22×10^{-5}	1.41×10^{-5}	3.50×10^{-1}
2000	2.62×10^{9}	2.72×10^{-5}	1.28×10^{-5}	9.08×10^{-1}

for Z(A), Z(A_2), Λ(A), and $\Lambda(A_2)$, it is easy to show that K_t is proportional to (see Workbook SM11.2, Cell 3)

$$\frac{m^{3/2}T^{5/2}}{p}$$

where m is the mass of the atom (H for H_2, and I for I_2). The difference in the values of K_t for I_2 and H_2 reflects the large difference in the mass of the atoms. In both cases K_t is much larger than one. This means that *translational motion always favors dissociation.*

§16. *Vibrational Contribution.* To calculate the contribution K_v of vibrational motion to the equilibrium constant, I must evaluate q_v (see Eq. 11.35). This is given by

$$q_v = \frac{\exp\left[D_0/k_B T\right]}{1 - \exp\left[T_v/T\right]} \tag{11.40}$$

and therefore

$$K_v = \frac{1}{q_v} = \exp\left[-\frac{D_0}{k_B T}\right]\left(1 - \exp\left[\frac{T_v}{T}\right]\right) \tag{11.41}$$

We encounter again the exponential $\exp[-D_0/k_B T]$ which played such an important role in the simple model discussed in §3 on p. 180. Because $T_0 \equiv D_0/k_B$ is much

Table 11.4 Vibrational parameters for H_2 and I_2.

Molecule	D_0 (eV)	$\hbar\omega(\text{cm}^{-1})$	T_v(K)
H_2	4.48	4401	6215
I_2	1.54	214	308

larger than T (see Table 9.1), K_v is extremely small. As you will see shortly, the fact that $\exp[-D_0/k_B T]$ is very small dominates the order of magnitude of the equilibrium constant, as it did in the simple model.

Workbook

§17. *Numerical Results for the Vibrational Contribution.* The parameters needed for evaluating K_v for H_2 and I_2 are given in Table 11.4. Using Eq. 11.41, I calculated (in Workbook SM11.3) the values of K_v listed in Tables 11.2 and 11.3. Even at high temperatures (e.g. 2000 K) the vibrational contribution is very small. If only the vibrational motion mattered, the concentration of atoms in the system would be very low. Note also that K_v for I_2 is much larger than for H_2, because D_0 for H_2 is larger than for I_2. The larger D_0 is, the smaller K_v becomes and fewer atoms will be present in the gas at equilibrium.

§18. *The Role of Rotational Motion.* The rotational contribution K_r to the dissociation equilibrium constant can be calculated from

$$K_r = \frac{1}{q_r} = \frac{2T_r}{T} = \frac{2B}{k_B T} \tag{11.42}$$

I have used here the equation $q_r = T/2T_r$ for the rotational partition function and $T_r = B/k_B$ for the rotational temperature. The factor of 2 appears because H_2 and I_2 are homonuclear molecules (have identical nuclei). If the atoms in the diatomic are different, $K_r = T_r/T$. Eq. 11.42 shows that the higher the rotational constant (or the rotational temperature), the larger the rotational contribution to the equilibrium constant K. In all cases, the rotation hinders dissociation since K_r is less than 1; the presence of rotation makes K smaller.

The factor 2 in K_r has some intriguing consequences. This factor appears because of quantum mechanical reasons that I explained in Chapter 9 and especially in Supplement 9.1. It must be there if both nuclei in the diatomic are the same (e.g. I_2)

but not if they differ (e.g. HI). Consider a molecule such as $^{17}O^{16}O$ in which the nuclei are different: how does its dissociation equilibrium constant compare to that for $^{16}O^{16}O$? The atoms ^{17}O and ^{16}O differ only in the presence of an additional neutron in the nucleus of ^{17}O. This does not affect the potential energies of the two molecules, because those are determined by the electrons, and the electrons do not interact with the extra neutrons. Therefore, the dissociation energies of the two molecules are nearly equal. Because the masses of $^{17}O^{16}O$ and $^{16}O^{16}O$ are close, their vibrational frequencies and rotational constants B are very close. You would think that their equilibrium constants ought to be practically the same. However, statistical mechanics says that this is not true. Because of the factor σ in the rotational partition function, the value of K for $^{16}O^{16}O$ dissociation is twice as big as that for $^{17}O^{16}O$ dissociation. This is a very counter-intuitive prediction, which could be verified experimentally. I don't know whether this prediction has been checked by experiments, but I am confident that if it tested, it will be found to be true.

§19. *The Total Equilibrium Constants.* I can now compare the correct statistical theory of dissociation equilibrium with the simplified two-level model discussed earlier. In spite if its crudity, the two-level system correctly predicts the most important feature of the dissociation equilibrium is the dissociation energy. The factor $\exp[-D_0/k_B T]$ appears in both the correct model (through K_v) and in the two-level one. For most molecules, this factor is so small that it controls the magnitude of the equilibrium constant. It is because of it that we must raise the temperature very high if we want to observe dissociation. There are, however, other factors. The translational contribution increases K substantially. The rotational one diminishes it. However, the favorable effect of translation always overcomes that of rotation.

Exercise 11.7

(a) Use statistical mechanics to calculate the equilibrium constant and the equilibrium composition for the reaction $H_2 + I_2 \rightleftharpoons 2HI$. Study the contribution of the translation, vibration, and rotation separately. What is the largest factor controlling the equilibrium concentration?

(b) Use thermodynamic methods to calculate the equilibrium constant of the reaction. Compare the result with the one obtained from statistical mechanics.

Exercise 11.8

The interaction energy between two Xe atoms is

$$V(r) = 4\epsilon\left[\left(\frac{a}{r}\right)^{12} - \left(\frac{a}{r}\right)^{6}\right]$$

with $\epsilon/k_B = 229$ K and $a = 4.055$ Å.

(a) For Xe_2, calculate the bond length r_e, the vibrational frequency ω, the rotational constant B, and the dissociation energy D_0.

(b) Calculate the fraction of Xe_2 molecules in a Xe gas, as a function of temperature.

(c) Compare the results to those obtained in the text for I_2. Why are the equilibrium constants so different?

Exercise 11.9

Compare the equilibrium constant for the reaction

$$H_2 + I_2 \rightleftharpoons 2HI$$

to that for the reaction

$$D_2 + I_2 \rightleftharpoons 2DI$$

D_2 and H_2 have the same force constant, the same bond length, and the same dissociation energy D (not D_0! — see Chapter 9, the section on the vibrational partition function).

(a) Compare the equilibrium constants numerically.

(b) Do a detailed analysis to show where the differences between the two equilibrium constants come from.

12

TRANSITION STATE THEORY: FUNDAMENTAL CONCEPTS

Introduction

§1. *A Review of Chemical Kinetics.* Chemical kinetics is a phenomenological theory, in which we guess a rate equation and then we solve it, to obtain an expression for the evolution of the concentration in time. If this expression fits the data, we are satisfied with our guess. If not, we start over: make a new guess, fit, etc.

This guesswork is guided by rules, based on experience, that depend on the type of reaction. For example, an irreversible isomerization reaction

$$\mathrm{A} \rightarrow \mathrm{B},$$

in which the compound A is transformed into B, is usually a first-order reaction. The rate equation for such cases is

$$\frac{dC_{\mathrm{A}}}{dt} = -k\, C_{\mathrm{A}}$$

Here C_{A} is the concentration of the reactant and t is time. The rate equations for all isomerization reactions, are the same; the only difference, from reaction to reaction, is the value of the *rate constant* k.

This equation has the solution

$$c_{\mathrm{A}}(t) = c_{\mathrm{A}}(0)\exp[-kt] \tag{12.1}$$

where $c_{\mathrm{A}}(0)$ is the initial concentration of compound A. The numerical value of k, for a specific reaction, is determined by forcing the concentration of the reactant at different times (given by Eq. 12.1) to fit the concentrations obtained by measurements.

A huge body of experimental data shows that the rate constant k of an elementary reaction satisfies, to a very good approximation, the Arrhenius equation

$$k = A\,\exp\left[-\frac{E_a}{k_B T}\right] \tag{12.2}$$

The *pre-exponential* A and the *activation energy* E_a are independent of the temperature T. k_B is the Boltzmann constant. A and E_a are found by forcing Eq. 12.2 to fit the values of the rate constant k at several temperatures.

This empirical procedure saves labor, but provides little illumination: its success raises a number of deeper questions, for which it has no answer. Why is there a rate law? What determines the form of the rate equation? What determines the order of magnitude of the rate constant? Where does the Arrhenius law come from? Why is it so universal? What is the meaning of the pre-exponential A and of the activation energy E_a? How does the solvent affect the rate constant? Where does the energy needed to cause a reaction come from? What is the qualitative picture that connects the rate of a reaction to the motion of the atoms participating in it?

In Chapters 12 to 15, I develop an approximate theory, called *transition state theory* (TST), that answers these questions.

§2. *A Brief Summary of the Topics Covered in Chapters 12 to 15.* In the present chapter I outline the ideas of the theory, to offer you a guide to the remaining chapters. You may not understand these ideas fully, but at least you will know where we are going. In Chapter 13 we study a simple model: the migration of an atom along a solid surface. This is used as an example, to explain and illustrate the theory. After this preparation, I derive the theory of the rate constant (Chapter 14). This is used in Chapter 15 to calculate the rate constant that describes the motion of an atom on the surface.

For the sake of simplicity, I limit the presentation to an ensemble of non-interacting molecules: each molecule reacts independently of the others. Moreover, I deal only with unimolecular reactions: those in which a molecule isomerizes or breaks into two fragments.

§3. *A Summary of What You Will Learn in this Chapter.* I begin this chapter by reviewing the concept of *potential energy surface* of a molecule (PES). This is a function $V(\mathbf{r}_1, \ldots, \mathbf{r}_N)$ that gives the potential energy of the molecule, when given the positions $(\mathbf{r}_1, \ldots, \mathbf{r}_N)$ of the N atoms in it. Chemical kinetics, chemical equilibrium, and the spectroscopic properties of a molecule all depend on the *shape* of the potential energy surface.

I will first argue that if a group of atoms form a molecule, the structure of that molecule corresponds to a minimum in the potential energy surface. For example, H, C, and N can form the molecules HCN or CNH. These two molecular structures correspond to two distinct minima of the potential energy surface.

In an isomerization reaction, which changes the molecule A into the molecule B, the atoms must move from a geometry corresponding to the minimum describing compound A, to one corresponding to the minimum describing the molecule B. The most difficult step in this trip is getting over the energy "ridge" separating these minima. The point on this ridge that has the smallest potential energy is called the *transition state* (TS). The difference between the energy of the transition state and that of the reactant is the activation energy. If you are a hiker it should not be hard to visualize what I am saying. Imagine that you start at campsite A, nestled at the bottom of a valley. Your destination is campsite B, which is at the bottom of the neighboring valley. The two valleys are separated by a ridge. The transition state is the lowest point on that ridge. A hiker will call that the "pass."

§4. *The Reaction Requires Energy and Takes it From the Environment.* On the potential energy surface, the transition state is a point of high energy. The molecule can cross the ridge only when it manages to "borrow" energy from its neighbors. When the temperature is low there isn't much energy in the system and the neighbors are tough bankers; only rarely do they give the molecule enough energy to react. It is this parsimony that makes a chemical reaction slow at low temperature. If the temperature is high, the neighbors have more energy and are more generous. The energy to cross the ridge is more readily available and the reaction speeds up.

Having more energy than the ridge is not enough. In the reaction

$$\mathrm{HCN} \rightleftharpoons \mathrm{CNH}$$

this energy has to be given to the hydrogen atom, which must swing around the C–N bond from the carbon end to the nitrogen end. Having a lot of energy in the C–N bond is not sufficient for promoting this reaction. This observation will lead us to the concept of *reaction path*: to react a molecule must gain enough energy in a specific kind of motion.

Finally, in this isomerization, once CNH is formed, the hydrogen atom must lose energy; if it doesn't, the energetic H atom might leave its position in CNH and go back to form HCN.

In what follows I will present arguments that clarify this picture and make it more plausible. I emphasize that the description given here is valid for *thermal reactions;* it does not hold for photochemical reactions or for those produced by molecular or ion-beam collisions, where the system is not in thermal equilibrium.

The Potential Energy Surface

§5. I am sure that you have already encountered the concept of potential energy surface of a molecule. This is so central to chemistry that I don't mind reviewing it again. If the molecule of interest has N atoms, then I denote by $\mathbf{r}_1$ the position of atom 1, by $\mathbf{r}_2$ the position of the atom 2, and so forth. I will call the list $C = \{\mathbf{r}_1, \mathbf{r}_2, \ldots, \mathbf{r}_N\}$ a *molecular configuration*. Each configuration defines a fixed spatial arrangement of the N atoms.

The potential energy surface $V(\mathbf{r}_1, \mathbf{r}_2, \ldots, \mathbf{r}_N) \equiv V(C)$ gives the potential energy of the molecule when the atoms have the configuration $C = \{\mathbf{r}_1, \mathbf{r}_2, \ldots, \mathbf{r}_N\}$. If I change the configuration C, the distance between some of the atoms or the angle between some of the bonds change, and the potential energy $V(C)$ changes.

I note in passing that a change of configuration that represents a translation or a rotation of the molecule as a whole does not change the potential energy.

§6. *A Chemical Compound is a Minimum on the Potential Energy Surface.* The potential energy surface of a molecule is complicated and, being a function of many coordinates, cannot be visualized. Fortunately, the potential energy surfaces of all molecules share a number of *qualitative features* that are of great importance to physical chemistry.

As a beginning chemist you have by now become accustomed with the fact that every molecule manages to maintain its atoms in well-defined positions. For example, the CO_2 molecule is linear and the oxygen–carbon distance is 1.1615 Å. This information defines a specific configuration, which we call the *structure* of the molecule. Most molecules have a unique structure, even though in principle their atoms could have an infinite number of configurations. If the molecule has an isomer, its atoms form two molecular structures, one for each isomer.

Exercise 12.1

Calculate the configurations specified when I tell you that CO_2 is linear and the C–O bond lengths are 1.1615 Å. (*Hint*. Place the carbon atom at the origin of the coordinate system O and the OZ axis along the molecule. Calculate the coordinates of the three atoms.)

In what follows I use statistical mechanics to argue that each molecular structure corresponds to a configuration for which the potential energy has a minimum.

I know that a molecule in a gas, or a liquid or a solid, collides very frequently with the other molecules. These collisions cause each molecule to gain or lose energy. When the molecule gains energy, the motion of its atoms is wilder and they can go through a large number of configurations. Harsh collisions can force the molecule into configurations that differ greatly from the molecular structure.

Should we conclude that the atoms in the molecule run all over the place, going through a sequence of very different configurations $C = \{\mathbf{r}_1, \mathbf{r}_2, \ldots, \mathbf{r}_N\}$? The answer is no: such a conclusion is in conflict with the facts. We know that most molecules are fairly rigid and their atoms mostly perform small oscillations around the structure of the molecule. If the atoms in the molecule were to perform very large excursions, all the time, the idea of a molecular structure would make no sense.

How does statistical mechanics explain the existence of molecular structure in spite of the fact that a molecule collides continuously with other molecules? And how can we determine what structure a molecule prefers?

In classical statistical mechanics, the probability that a molecule has a configuration $C = \{\mathbf{r}_1, \mathbf{r}_2, \ldots, \mathbf{r}_N\}$ is

$$p(C) = \frac{1}{Q} \exp\left[-\frac{V(C)}{k_B T}\right] = \frac{1}{Q} \exp\left[-\frac{V(\mathbf{r}_1, \mathbf{r}_2, \ldots, \mathbf{r}_N)}{k_B T}\right] \tag{12.3}$$

Here Q is the classical configurational partition function. All you need to know about Q is that it depends on temperature and density but not on the molecular configuration; when the molecule changes its configuration C, Q does not change.

From this formula it is clear that the most probable configuration C_{mp} is the one for which the function $V(\mathbf{r}_1, \mathbf{r}_2, \ldots, \mathbf{r}_N)$ has a minimum: the configuration C for which $V(C)$ is smallest has the highest probability.

Here is how we interpret this result. Let us examine a gas with a large number N of molecules. Denote by $\mathcal{N}(C)$ the number of molecules whose configuration is very close to C. Don't worry what "very close" means; for what I do here, it is not essential to have a precise definition. "Very close" is whatever you think that it should be. Eq. 12.3 tells me that

$$\frac{\mathcal{N}(C)}{\mathcal{N}(C_{\text{mp}})} = \exp\left[-\frac{V(C) - V(C_{\text{mp}})}{k_B T}\right] \tag{12.4}$$

Because $V(C_{\text{mp}})$ is the minimum potential energy, $V(C) - V(C_{\text{mp}})$ is positive and the ratio $\mathcal{N}(C)/\mathcal{N}(C_{\text{mp}})$ is smaller than 1, no matter what the configuration C is configuration is "very close" to C_{mp} are most frequently encountered in the system. That is, *most molecules* will take a configuration that minimizes their potential energy. This, most frequently encountered, configuration, is the structure of the molecule.

It is a fact of life that most molecules have a well defined structure (polymers are a notable exception). Eq. 12.4 explains why this happens: the change $\Delta V \equiv V(C_{\text{mp}} + \delta C) - V(C_{\text{mp}})$ of the potential energy, when the molecular configuration changes from C_{mp} to $C_{\text{mp}} + \delta C$, is larger than $k_B T$ even when δC is small. Small changes in geometry cause large changes in the potential energy (as compared to $k_B T$). This is why in experiments, the geometry of a very large fraction of molecules is very close to C_{mp}.

For many polymers, there are many large changes in molecular configuration that cause a small change in potential energy. This means that these configurations, which are rather different from C_{mp}, are frequently present in the system.

Exercise 12.2

The potential energy "surface" for a diatomic is often represented by the Morse potential

$$V(r) = -D + D\left[1 - \exp(-\alpha(r - r_0))\right]^2$$

For CO, $D = 11.2254$ eV, $r_0 = 1.1283$ Å, and $\alpha = 3.2145$ Å^{-1}.

(a) Calculate the molecular structure of CO (in this case, the bond length).

(b) Calculate the ratio between the number of molecules whose bond length is 1.17 Å and the number of molecules whose bond length is that corresponding to the molecular structure.

(c) Plot the probability that the bond length has a given value r. (*Hint.* Use Eq. 12.3 normalized so that $\int_0^\infty p(r)dr = 1$. Plot $p(r)$ versus r.)

---

§7. *Most of the Time a Molecule's Configuration is Close to* C_{mp}. There is an alternative interpretation of Eq. 12.4 which is equally valid and is often very useful. Let us assume that I observe, for a very long time τ, the configuration of one molecule in the system. In the case of a diatomic this means that I observe what happens to the bond length. For a triatomic, I monitor the two bond lengths and the bond angle. Because the atoms in the molecule have some kinetic energy, due to collisions with the other molecules, its configuration changes with time.

Let me denote by $\tau(C)$ the total time that the molecule spends close to the configuration C during the observation time τ. We have

$$\frac{\tau(C)}{\tau(C')} = \frac{\mathcal{N}(C)}{\mathcal{N}(C')} = \exp\left[-\frac{V(C) - V(C')}{k_B T}\right] \tag{12.5}$$

The first equality is a postulate of statistical mechanics.

This is not hard to understand. Imagine that you are in charge of a national park and you want to get an idea of its utilization. You have five attractions, labeled A_1, $A_2, \ldots, A_5$. You can count how many tourists are at A_1, how many at A_2, etc., at a given time. But you can also follow one tourist and note how much time he spent at A_1, how much time at A_2, etc. You get the same information, if all the tourists are identical. Of course the tourists aren't identical, but the molecules are and

$$\frac{\tau_1}{\sum_i \tau_i} = \frac{N_1}{\sum_i N_i}$$

where τ_i is the time spent at A_i and N_i is the number of tourists at A_i.

Eq. 12.5 leads me to several important conclusions. A molecule spends the longest time in the lowest energy configuration C_{mp} and it spends very little time in a configuration C for which

$$\frac{V(C) - V(C_{\text{mp}})}{k_B T} \gg 1$$

This means that a molecule having high energy is most likely to lose it rapidly, because it can only spend a very short time in that state. The higher the potential energy of a configuration, the shorter the time the molecule spends in that configuration.

However, since my conclusions are derived from equations involving probabilities, they are not absolute. Occasionally, a molecule in a configuration with high potential energy may evolve to a configuration of even higher energy. Such events are rare, but they are the events leading to chemical reactions.

§8. *Atoms Having High Kinetic Energy Tend to Lose it.* The way the atoms in a molecule move depends on both their potential energy and their kinetic energy. So far, I have shown that in an ensemble of molecules (gas, liquid, ...), the atoms in a molecule tend to spend more time in a configuration having low potential energy. If they accidentally happen to have high potential energy, they will tend to lose it rapidly.

The same can be said about the kinetic energy. The argument is very similar to the one made for the potential energy. The velocities of the N atoms in a molecule are described by the list $\mathcal{V} = \{\mathbf{v}_1, \mathbf{v}_2, \ldots, \mathbf{v}_N\}$, where $\mathbf{v}_1$ is the velocity of atom 1, etc. The probability that the atoms have velocities in the neighborhood of $\mathcal{V}$ is given by the Maxwell velocity distribution

$$D \exp\left[-\frac{\mathcal{K}(\mathcal{V})}{k_B T}\right] \tag{12.6}$$

Here

$$\mathcal{K}(\mathcal{V}) = \frac{m_1\mathbf{v}_1^2 + m_2\mathbf{v}_2^2 + \ldots + m_N\mathbf{v}_N^2}{2}$$

is the total kinetic energy of the molecule, m_1 is the mass of atom one, m_2 is that of atom 2, ... , and D is independent of the velocities.

Eq. 12.6 is very similar to Eq. 12.3 and can be interpreted in the same way. I denote by $\mathcal{N}(\mathcal{V})$ the number of molecules whose kinetic energy is in the neighborhood of $\mathcal{V}$. The ratio $\mathcal{N}(\mathcal{V})/\mathcal{N}(\mathcal{V}')$ is equal to the ratio of the respective probabilities $\mathcal{P}(\mathcal{V})$ and $\mathcal{P}(\mathcal{V}')$:

$$\frac{\mathcal{N}(\mathcal{V})}{\mathcal{N}(\mathcal{V}')} = \frac{\mathcal{P}(\mathcal{V})}{\mathcal{P}(\mathcal{V}')} = \exp\left[-\frac{\mathcal{K}(\mathcal{V}) - \mathcal{K}(\mathcal{V}')}{k_B T}\right] \tag{12.7}$$

The last equality follows from Eq. 12.6. Using Eq. 12.7, I find that if

$$\mathcal{K}(\mathcal{V}) > \mathcal{K}(\mathcal{V}')$$

then

$$\mathcal{N}(\mathcal{K}(\mathcal{V})) < \mathcal{N}(\mathcal{K}(\mathcal{V}')).$$

It is less likely to find molecules whose atoms have high kinetic energy than to find molecules with sluggish atoms. Put more precisely, the fraction of molecules in which the kinetic energy of the atoms exceeds the energy $k_B T$ is very small. Increasing the temperature increases the probability of finding molecules whose atoms have a high kinetic energy.

If I watch one molecule for a long time, I will find (by an argument similar to that made for potential energy) that a molecule spends very little time in a state with kinetic energy much larger than $k_B T$. If it happens to aquire such a high kinetic energy, because of a particularly violent collision with another molecule, it will lose it very rapidly.

§9. *The Potential Energy Surface of a Molecule with Two Isomers.* Let us consider, as an example, a triatomic molecule that can undergo a rearrangement (isomerization) reaction, such as

$$\mathrm{HCN} \rightleftharpoons \mathrm{CNH}.$$

The two chemical compounds HCN and CNH have different structures, corresponding to two different molecular configurations $C_{\mathrm{HCN}} = \{\mathbf{r}_{\mathrm{H}}^{(1)}, \mathbf{r}_{\mathrm{C}}^{(1)}, \mathbf{r}_{\mathrm{N}}^{(1)}\}$ and $C_{\mathrm{CNH}} = \{\mathbf{r}_{\mathrm{H}}^{(2)}, \mathbf{r}_{\mathrm{C}}^{(2)}, \mathbf{r}_{\mathrm{N}}^{(2)}\}$. The only difference between these molecules is in the positions of the atoms: the numerical values of $\mathbf{r}_{\mathrm{H}}^{(1)}, \mathbf{r}_{\mathrm{C}}^{(1)}, \mathbf{r}_{\mathrm{N}}^{(1)}$ differ from those of $\mathbf{r}_{\mathrm{H}}^{(2)}, \mathbf{r}_{\mathrm{C}}^{(2)}, \mathbf{r}_{\mathrm{N}}^{(2)}$.

We know from experiments that both HCN and CNH are stable molecules: each has a well defined structure. This means that the configurations C_{HCN} and C_{CNH} must correspond to minima on the potential energy surface of the system formed by the three atoms (i.e. C, N, and H). From Chapter 11, on chemical equilibrium, we know that if $V(C_{\mathrm{HCN}}) < V(C_{\mathrm{CNH}})$ then HCN is more stable than CNH. This also follows from Eq. 12.3. The fact that HCN is more stable does not mean that the solution contains only HCN; it means that, at equilibrium, the solution contains more HCN than CNH. The equilibrium concentration can be calculated with the theory explained in Chapter 11.

The conclusion reached in this example is general. If a group of atoms can form several isomers, the structure of the isomers corresponds to minima on the potential energy surface of that particular group of atoms. When the isomers are in equilibrium, the one with the lowest potential energy has the highest concentration.

§10. *Between Two Minima There Must be a Ridge.* I will appeal now to your geometric intuition and to your experience as a hiker (hiking in the mountains is a good way to study surfaces). Let us look at a system with two degrees of freedom, so that you can visualize what is going on. This means that the potential energy $V(x, y)$ depends on only two variables, x and y. Never mind what the system is and what x and y are: we are doing mathematics now. I hope that you remember, from your study of calculus, that the function $z = V(x, y)$ represents a surface in the three-dimensional space $\{x, y, z\}$. Let us assume that the system has two isomers $\mathcal{A}$ and $\mathcal{B}$. This means that the surface has two minima, one corresponding to the structure of $\mathcal{A}$ and the other to the structure of $\mathcal{B}$. Try to imagine these minima as two bowls on the surface.

One way of representing such a surface (i.e. the function $V(x, y)$) is to make a contour plot (see Fig. 12.1). The topographic maps used by hikers have such plots on them; the contours join points having the same height. In our case, the contours join points having the same potential energy. The minima on this surface are indicated by the letters A and B. They are separated by a ridge shown by the line joining M to N. It is an accident that the ridge is a straight line; there is no reason to expect this in general. The arrows in Fig. 12.1 indicate the uphill directions on the surface.

Let us now look at the properties of the ridge. To reach the ridge from either A or B, one has to climb uphill. Therefore, a hiker moving along the x direction will find that the ridge corresponds to the maximum height in his trip. The ridge is the locus of all these maxima. If you travel along the ridge (say from M to N) you will find a point where the ridge has the lowest height. This point is called *the transition state* (TS), and it is marked by a solid circle in Fig. 12.1.

If you travel away from the TS, towards the minima, you go downhill; if you travel, away from the TS, along the ridge, you go uphill. The region around the TS resembles the region around the center of a saddle. Because of this, the transition state is often called *the saddle point.*

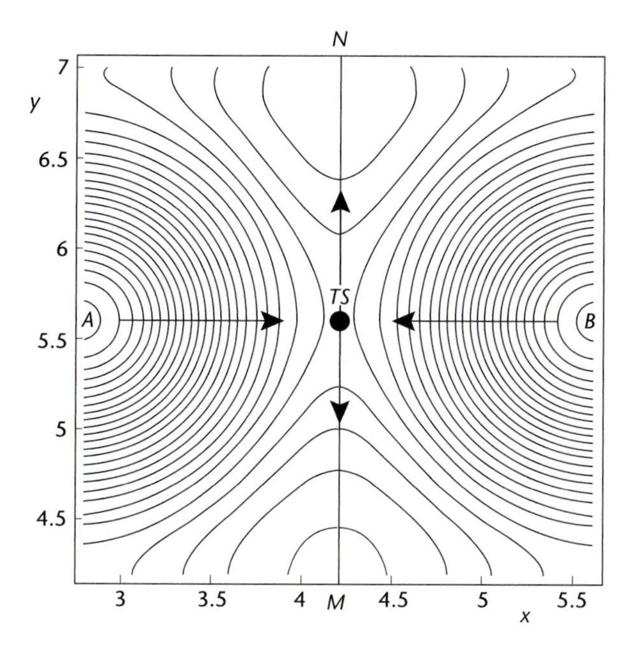

Figure 12.1 Contour plot of a two-dimensional potential energy surface $V(x, y)$ that has two minima: one at A and another at B. The ridge is along the line joining M to N. The solid circle is the transition state. The arrows point in the uphill direction on the surface.

If contour plots intimidate you, take a look at Fig. 12.2, which shows a regular, three-dimensional plot of the same surface. You can see the saddle shape around the transition state and the two minima.

§11. *What Does This Have to do with Chemical Kinetics?* The existence of this ridge, and of the transition state on it, are the two most important concepts in chemical kinetics. Here is why.

Let us look at the isomerization reaction

$$\mathcal{A} \to \mathcal{B}.$$

The potential energy surface (PES) of this system is described by Fig. 12.1 (or Fig. 12.2). When the system forms compound $\mathcal{A}$, its atoms have configurations $C(\mathcal{A})$, which are all close to the minimum A of the PES. At equilibrium,

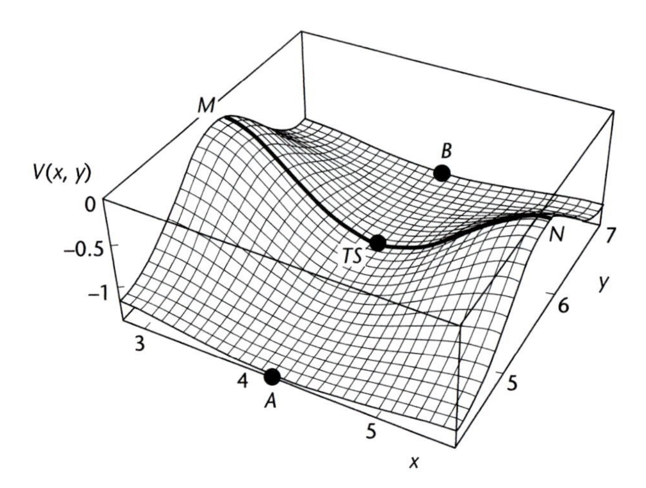

Figure 12.2 A plot of the surface $V(x, y)$ shown in Fig. 12.1. The minima are at A and B. The ridge is along the line joining M to N. The solid circle, in the middle of the figure, is the transition state.

some molecules have configurations $C(\mathcal{B})$, located in the neighborhood of minimum B in Fig. 12.1.

Imagine now that we prepare a system that contains only $\mathcal{A}$ molecules. The system will evolve to equilibrium by converting some of the $\mathcal{A}$ molecules into $\mathcal{B}$. For this to happen, the configuration of the $\mathcal{A}$ molecules must evolve from $C(\mathcal{A})$ to $C(\mathcal{B})$. There must be a *reactive trajectory* that starts in $C(\mathcal{A})$ and ends in $C(\mathcal{B})$ (see Fig. 12.3). Along this trajectory, the system must cross the ridge separating A from B.

The energy of a molecule whose configuration is on the ridge is higher than the energy of either minimum. According to Eq. 12.3, the probability that a molecule reaches such a high energy is low; the configuration of most molecules in the system is close to the points A and B in Figs 12.1 or 12.2.

I denote now the configuration in the neighborhood of the transition state by $C(\mathrm{TS})$ and that near A (corresponding to the compound $\mathcal{A}$) by $C(\mathrm{A})$. Consider now an ensemble of molecules $\mathcal{A}$, and denote by $\mathcal{N}(C(\mathrm{A}))$ the number of molecules whose configuration is close to $C(\mathrm{A})$, and by $\mathcal{N}(C(\mathrm{TS}))$ the number of molecules whose configuration is close to $C(\mathrm{TS})$. By using Eq. 12.3 we find that

$$\frac{\mathcal{N}(C(\mathrm{TS}))}{\mathcal{N}(C(\mathrm{A}))} = \exp\left[-\frac{\mathrm{V(TS)} - \mathrm{V(A)}}{k_B T}\right] \tag{12.8}$$

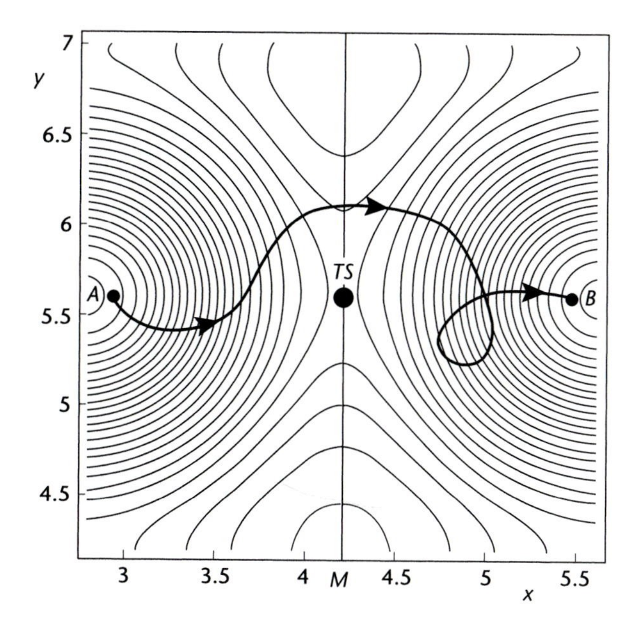

Figure 12.3 A reactive trajectory that transforms molecule $\mathcal{A}$ into molecule $\mathcal{B}$.

In this equation and the discussion that follows I use the notation $V(\mathrm{A})$ for $V(C(\mathrm{A}))$ and $V(\mathrm{TS})$ for $V(C(\mathrm{TS}))$. We know that the difference

$$V(\mathrm{TS}) - V(\mathrm{A})$$

is positive, since the transition-state energy $V(\mathrm{TS})$ is higher than the energy $V(\mathrm{A})$ of the compound $\mathcal{A}$. If

$$V(\mathrm{TS}) - V(\mathrm{A}) \gg k_B T \tag{12.9}$$

then an extremely small fraction of the molecules in the ensemble will have a configuration close to the transition state.

Exercise 12.3

The energy difference between the transition state and that of the reactant $\mathcal{A}$ is 0.6 eV.

(a) Calculate the fraction of molecules whose configuration is close to that of the transition state.

(b) Perform the same calculation if the energy difference is 90 kcal/mol.

This conclusion has immediate implications for kinetics. During the reaction the point representing the molecular configuration has to pass over the ridge. Let us denote by C_r any configuration on the ridge. If $\mathcal{N}(C_r)/\mathcal{N}(A)$ is very small, only very few molecules reach the ridge. This means that only a few molecules react (go over the ridge, into the minimum corresponding to $\mathcal{B}$).

Of all the points on the ridge, *the transition state has the lowest energy*. This means that among the molecules whose configuration is close to the ridge, most are near the transition state; the crossing of the ridge is most likely to take place near the transition state. The transition state is like a mountain pass on a trail: if a choice is available, most people getting into the mountains are likely to go through the lowest mountain pass. While for hikers this is optional (a matter of common sense), for molecules this is dictated by the laws of statistical mechanics.

At a given temperature, the higher the difference V(TS) − V(A), the smaller the number of molecules reaching the transition state (see Eq. 12.8) and the smaller the number of molecules that react.

You have probably noticed the resemblance of the factor

$$\exp\left[-\frac{V(TS) - V(A)}{k_B T}\right]$$

(which appears in Eq. 12.8) to the factor

$$\exp\left[-\frac{E_a}{k_B T}\right]$$

that appears in the Arrhenius equation (Eq. 12.2). If we assume that the rate constant is proportional to the number of molecules that can reach the transition state, then we are led to conclude that

$$E_a = V(TS) - V(A) \tag{12.10}$$

This is a breathtaking conclusion: with qualitative, careful thinking, we have come to identify the meaning of the activation energy: it is the difference between the energy of the saddle point (the transition state) and that of the reactant. Our argument is not a proof but it is a powerful conjecture. A complete proof will be given in the chapters that follow. There you will find that Eq. 12.10 is not exact but is a very good approximation.

§12. *Summary of the Conclusions Reached in the First Stage of the Argument.* If you study a molecule having N atoms that can form two isomers, $\mathcal{A}$ and $\mathcal{B}$, then the potential energy surface of those atoms has minima corresponding to the structures of $\mathcal{A}$ and $\mathcal{B}$.

Statistical mechanics tells me that the fraction of molecules whose kinetic or potential energy substantially exceeds $k_B T$ is very small. The few that manage to have such high energy tend to lose it very rapidly. As a consequence, the molecular configuration of most molecules is close to the one that minimizes the potential energy (i.e. the molecular structure) and the kinetic energy of the atoms in the molecule is of order $k_B T$. For the temperatures used ordinarily in chemistry, $k_B T$ is of order of 0.025 eV, which is much smaller than the strength of a chemical bond (which could be between 0.4 eV and 11 eV).

We conclude that atoms in molecules are sedentary. You can imagine them spending most of their time at home (the zone of highest comfort) and rarely having enough energy to get out and visit a remote place. Those who manage to acquire sufficient energy are the ones that engage in chemical reactions. This is why we often say that a chemical reaction is a *rare* event, since it is rare that a molecule has sufficient energy to react. A slow reaction is not really slow. When it happens, the system goes over the saddle point very quickly. The rate of the reaction is small because such transitions take place very rarely.

If the reaction $\mathcal{A} \to \mathcal{B}$ is possible, then the two minima (corresponding to $\mathcal{A}$ and $\mathcal{B}$) are adjacent and are separated by a ridge. To go from the minimum corresponding to $\mathcal{A}$ to that corresponding to $\mathcal{B}$, the system must get over the ridge. The most accessible point on the ridge is the transition state TS. The probability that the point TS is reached depends exponentially on the magnitude of $V(\mathrm{TS}) - V(\mathrm{A})/k_B T$. The larger this quantity, the smaller the reaction rate.

The Dynamics of a Chemical Reaction

§13. *To React, the Molecule Must Cross an Energy Ridge.* If a molecule has two isomers, then it can undergo an isomerization reaction. What can we say about the atomic motion causing the reaction to take place? Let us look at the specific example mentioned in §9. We start an experiment with a gas containing HCN and heat it up. We will observe that some HCN molecules disappear and some CNH

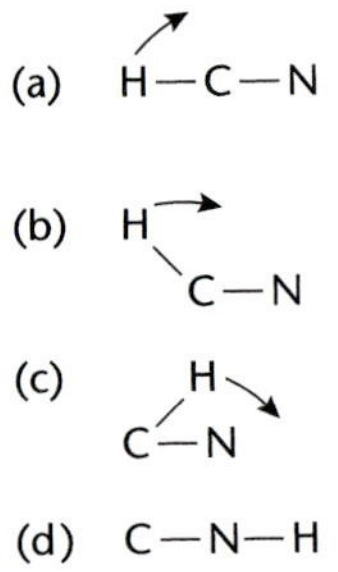

(a) H—C—N

(b) H C—N

(c) H C—N

(d) C—N—H

Figure 12.4 A sequence of configuration changes along a reactive trajectory evolving from HCN to CNH.

appears; the reaction HCN $\rightarrow$ CNH occurs in the system, and proceeds until equilibrium is reached. During the reaction the molecular configuration of many molecules changes from C_{HCN} to C_{CNH}. To produce such a change the atoms must move around. This motion is described by a trajectory that starts in the potential energy bowl corresponding to C_{HCN} and reaches the bowl corresponding to C_{CNH} (a *reactive trajectory*).

In Fig. 12.4, I show a few intermediate configurations along a reactive trajectory. For the reaction to take place the hydrogen atom has to swing around the CN group, from the carbon to the nitrogen. The C–N bond length will also change in the process, but not by a large amount.

§14. *The Energy Causing a Reaction Comes from the Medium.* Why would a molecule perform a reactive trajectory? Where does the energy required to go over the ridge come from? The short answer is: collisions. A molecule in a gas (or liquid or solid) is hit extremely often by its neighbors; it is a rough world out there. Because of this, the potential energy of the molecule and the kinetic energy of its atoms goes up and down all the time. Occasionally, and accidentally, the hydrogen atom acquires enough energy to move along a trajectory like the one shown in Fig. 12.4. That trajectory gets to the final state C–N–H only if the H atom in the molecule has enough energy to get over the ridge separating the two minima.

§15. *To Cause a Reaction, the Energy Must Go into the Right Kind of Motion: the Reaction Coordinate.* If you look at the snapshots in Fig. 12.4, it is clear that it is not enough that the *total energy* of the molecule is larger than that of the ridge separating HCN from CNH. The energy must be used to move the hydrogen atom, as shown in Fig. 12.4. If the collision energy goes into the C–N bond and makes it stretch, this will not cause the desired reaction; the hydrogen atom may be completely unimpressed by the fact that the C–N bond oscillates with a large amplitude.

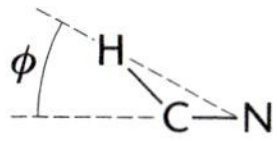

Figure 12.5 This figure shows that for a gross description of the progress of reaction, I can use the angle ϕ.

ϕ H C—N

This observation leads us to the concept of *reaction coordinate*. For our example, the reaction coordinate is (roughly) the angle ϕ shown in Fig. 12.5. If ϕ is zero, the molecule is HCN. If ϕ is 180°, the molecule is CNH. The molecule evolves from HCN to CNH only if the hydrogen atom moves in a way that increases ϕ from 0 to 180°.

The reaction coordinate is different for each reaction. For example, for

H(X)C=C(H)Y → H(X)C=C(Y)H

the reaction coordinate is the angle made by the plane defined by

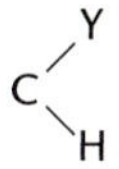

C(Y)H

with the plane defined by

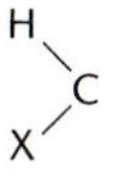

H(X)C

For the reaction $CH_3CH_3 \rightarrow 2CH_3$, the reaction coordinate is the carbon–carbon distance.

As you will see later, these definitions are oversimplified, but they display the essence of the concept.

§16. *Put it all Together.* Now you know what a chemical reaction is, in terms of the motion of the atoms comprising the molecule. The configuration representing the reactants starts its odyssey at the bottom of a bowl on the potential energy surface. The molecule is hammered by collisions with the atoms of the medium and its energy fluctuates. Occasionally, and rather rarely, the molecule acquires enough energy, *in the reaction coordinate*, to go over the ridge separating the reactant minimum from the product minimum. The lowest energy point on the ridge, called the transition state, is the point most likely visited by a reactive trajectory.

Can we turn this picture into a theory? All I have to do is to calculate the probability that the molecule reaches the ridge and the velocity with which is passes over it. These two quantities are given to me by statistical mechanics. If I am clever enough I ought to be able to build a theory of the rate equation (and the rate constant) from these two elements. In the next two chapters, I show how this is done.

13

TRANSITION STATE THEORY: THE MOTION OF A CHEMISORBED ATOM ALONG THE SURFACE

A Description of a Chemical Reaction

§1. *Introduction.* Practically all interesting reactions involve a change in the position of several atoms. To describe such a change, I need a large number of coordinates, and the potential energy surface is a function of many variables. Having lived in a three-dimensional space, we have trouble visualizing a surface in a space with many dimensions. By using abstraction, mathematics allows us to generalize our three-dimensional insights. However, since most of you are likely to have little experience with abstract spaces with many dimensions, I prefer to take the low road. I will develop the theory of the rate constant for a very simple example: the motion of an atom *chemisorbed* on a solid surface.

A chemisorbed atom (an atom that forms a chemical bond with a solid surface) prefers to bind to *specific sites* on the surface and moves around by jumping from binding site to binding site. This is the simplest "isomerization reaction" I could think of. Within a rough model, which assumes that the motion of the solid's atoms is of secondary importance, I can describe the mechanics of this reaction by following the trajectory of the chemisorbed atom. The potential energy surface will depend then only on the three coordinates of the chemisorbed atom. Generalizing, from this system to more complex ones is only a matter of mathematical abstraction; no new physical or chemical concepts need to be invoked.

In this chapter, I will calculate the potential energy surface for this simple system, determine the binding sites, and study the geometry of the potential energy surface around the binding sites. Then, I will find the ridge separating the binding sites and the transition state on it. I will also study the geometry of the potential energy surface around the transition state.

This is preliminary material needed as a starting point for developing the rate theory in Chapter 14.

§2. *The Structure of the System.* The atom is chemisorbed on a perfectly flat surface of a mono-atomic crystal. The positions of the atoms in the top layer of the solid surface are shown by empty circles in Fig. 13.1. Notice that they form a square lattice: each atom has four equidistant, nearest neighbors in the plane of the surface. The square shown by solid lines in Fig. 13.1 is called a *unit cell* of the square lattice. Another unit cell is shown by the dashed lines. I will tell you more about the positions of the atoms in the layers below the surface, when you need this information.

From calculations and experiments, I know that, often, an atom chemisorbed on a solid surface having square symmetry binds in the "four-fold" position (the solid circle in Fig. 13.1), which is at the center of the square unit cell. Other binding positions are possible, depending on the chemistry taking place between the adsorbed atom and the surface. I will consider here only the case of four-fold binding.

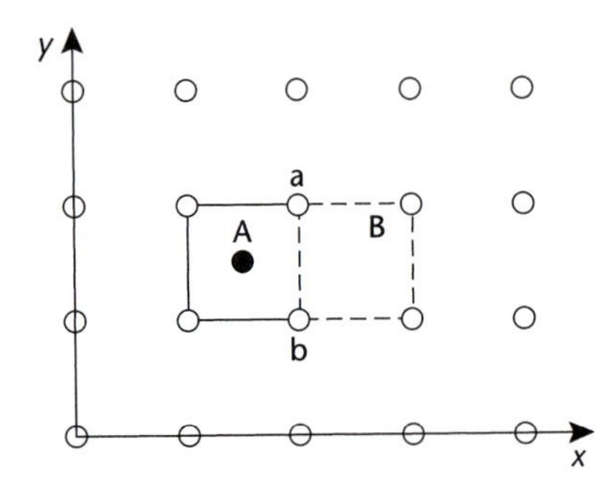

Figure 13.1 The empty circles show the positions of the atoms in the top surface layer of the solid. The black circle is the chemisorbed atom, binding to a four-fold site.

§3. *The Dynamics of the Adsorbed Atom.* To prepare the system I am going to study, I expose the solid surface to a gas of atoms that make a chemical bond with it. The concentration of the adsorbed atoms on the surface is very low and they do not interact with each other; no adsorbed atom cares what the other adsorbed atoms do. They have independent lives and move on the surface as if they are alone on it.

Let us assume that my vision is so good that I can follow the motion of individual atoms on the surface. What would I see? The surface atoms perform small-amplitude oscillations around the lattice sites. The chemisorbed atoms perform small oscillations around the four-fold site to which they like to bind. The amplitudes of these oscillations increase with temperature. The hotter the surface, the more agitated the atoms.

If I lack patience and I observe one chemisorbed atom for a short time, I will conclude that there is no tourism on the surface. Each adsorbed atom seems quite happy to stay in one unit cell.

My short-time observations are correct, but highly misleading. Had I observed the surface longer, I would have seen that the adsorbed atoms are avid travelers. However, like most of us, they travel while on vacation. If the observation time is much shorter than the time between vacations, I am likely to conclude that the atom does not leave the unit cell in which it resides. However, a more patient observer will find that, from time to time, the chemisorbed atom moves out of its cell of residence to one of the neighboring cells. For example, the atom shown in Fig. 13.1, which resides initially in cell A, will move, after some time, to cell B. As time goes on, each adsorbed atom travels along the surface, at its own slow pace, by jumping from one cell (binding site) to another.

§4. *How Fast is Fast for an Adsorbed Atom?* At this point you may ask: what do you mean by "slow" or "fast," when you are talking about chemisorbed atoms? From spectroscopic measurements or calculations, we know that the frequency of the oscillatory motion of the adsorbed atoms is on the order of 10^{14} s^{-1}; the atom performs a complete oscillation in about 10^{-14} s. The time an atom spends at a binding site depends very much on how strongly it binds to the surface and on temperature. It can vary from seconds to 10^{-10} s. The lifetime of the atom on a site is always considerably longer than the time in which the atom performs an oscillation around the binding site.

§5. *How Do I Know All This?* In the past decade microscopy has been improved so much that, in favorable cases, we can obtain a picture of one atom sitting on a solid surface. This can be done with a field ion microscope or with a scanning tunneling microscope. Here is how these instruments are used to learn about the motion of an atom on a solid surface. It takes a minute or so to obtain the picture of one atom, with one of these instruments. The picture cannot be taken if the atoms moves during this time. To keep the atoms still, we cool the surface to a very low temperature (about 20 K or 70 K, depending on the type of cooling used), to make sure that the atom spends more than a minute in the same unit cell. When the temperature is this low, we can take the picture and determine in which unit cell the atom resides. Next, we rapidly heat the surface to a constant temperature T and keep it at this temperature for a time τ. Then, we cool it rapidly to a very low temperature and take a picture of the surface, to see how far the adsorbed atom moved. From the information provided by such measurements one can extract the mean time the atom spends on a given lattice site.

We can also observe consequences of atomic motion. We can take pictures of a surface, at high temperature, during the deposition of atoms from the vapor phase. Now we cannot see individual atoms, because they move faster than the time in which a picture can be taken. However, as the atoms move around, they meet and form dimers, trimers, and larger clusters. These clusters are immobile (or move very slowly) and the instrument can take pictures of them, at different times during deposition. The pictures show that the clusters grow, as they trap more of the atoms moving on the surface.

If we stop the deposition, then we can observe that the clusters change shape because the atoms move around their border. Many other interesting phenomena, all caused by the site-to-site motion of the atoms on the surface, can be observed during such experiments.

There is no doubt that the atoms chemisorbed on the surface of a solid are constantly in motion and the topology of the surface is changing continuously. All of this happens because the atoms move from binding site to binding site.

§6. *Who Cares?* The motion of an adsorbed atom on a surface is scientifically interesting and deserves to be studied and understood. It is also important for understanding and controlling *epitaxy*, a process by which a variety of layered solid structures are made, by exposing a solid surface to the vapors of various solid materials. The solid state lasers used in the supermarket scanners, in your CD players, and in telecommunications are made this way. In getting the structure we want, we use very carefully the mobility of the atoms on the surface. Too little atomic mobility leads to rough layers which are unusable. If the mobility is too high, the atoms in one layer can travel into another layer, giving the wrong composition and poor properties.

The motion of atoms along a solid surface causes important phenomena to take place in catalytic systems. Many catalysts consist of small metal clusters stuck on an oxide surface. It is essential to disperse the metal as finely as possible, to achieve a very large area, because only the atoms at the surface of the cluster are catalytically active. When the catalyst is used, often at high temperature, the metal atoms move around the oxide surface and travel from one cluster to another. In this process there is a tendency for the small clusters to lose atoms and for the large clusters to gain them. As a result the metal clusters become larger and larger — this process is called coarsening. Coarsening lowers the total area of the catalyst and after a while (in about a year) the catalyst loses efficiency and needs to be replaced. This is most often the reason why the catalytic converter in your car needs changing. Every chemist designing a new catalytic process needs to worry about coarsening; to prevent it the mobility of the atoms on the oxide surface must be diminished.

The Potential Energy Surface of this System

§7. *The Energy of Interaction Between Two Atoms.* Since I plan to perform calculations, I need to have an equation for the potential energy surface of this system. The potential energy $V(\mathbf{r}, \mathbf{r}_1, \mathbf{r}_2, \ldots, \mathbf{r}_n)$ depends on the positions $\{\mathbf{r}, \mathbf{r}_1, \ldots, \mathbf{r}_n\}$ of all the atoms in the system. Here $\{\mathbf{r}_1, \mathbf{r}_2, \ldots, \mathbf{r}_n\}$ are the positions of the atoms forming the solid and $\mathbf{r}$ is the position of the chemisorbed atom. However, to keep the presentation simple, I will pretend that the atoms of the solid do not move. This is not true, but it is not a terrible approximation since at reasonable temperatures, on the time scale on which the adsorbed atom jumps from one site to another, the surface atoms perform small oscillations around their equilibrium positions.

This approximation simplifies tremendously the potential energy surface: if $\mathbf{r}_1, \mathbf{r}_2, \ldots, \mathbf{r}_n$ are kept fixed, the potential energy surface $V(\mathbf{r}, \mathbf{r}_1, \mathbf{r}_2, \ldots, \mathbf{r}_n)$ depends only on $\mathbf{r}$. In what follows I will denote this *restricted potential energy surface* by $V(\mathbf{r})$.

I would not make this approximation if I intended to calculate accurately the rate constant for a site-to-site jump. First, even though the surface atoms move over a short distance, they do contribute to the potential energy and to the rate constant. Second, neglecting the motion of the atoms in the solid is inconsistent with the general outline of the theory. If the atoms of the solid do not move, they cannot exchange energy with the adsorbed atom. This energy exchange is the source of the occasional energy excess that makes the chemisorbed atom jump to another site. This is like a cash infusion before a vacation: without it the atom does not travel and the reaction does not take place.

On the other hand, this approximation simplifies the presentation of the theory so greatly — by reducing the dimensionality of the problem — that I will make it anyway. Otherwise the presentation is complicated by unessential mathematics. You will see that we can take into account, approximately, the effect of the motion of the lattice atoms on the chemisorbed atom, without having to treat it explicitly in the potential energy surface. I will return to this approximation and explain how it can be removed, after I complete the presentation of the theory for the simplified model.

§8. *The Potential Energy Surface.* To calculate $V(\mathbf{r})$, I assume that the interaction energy between the atom at $\mathbf{r} \equiv \{x, y, z\}$ and an atom in the solid located at $\mathbf{r}_i \equiv \{x_i, y_i, z_i\}$ is

$$v(d_i) = 4\epsilon \left[\left(\frac{\sigma}{d_i} \right)^{12} - \left(\frac{\sigma}{d_i} \right)^{6} \right] \tag{13.1}$$

Here

$$d_i = \sqrt{(x - x_i)^2 + (y - y_i)^2 + (z - z_i)^2} \tag{13.2}$$

is the distance between the adsorbed atom and a solid atom located at $\mathbf{r}_i$.

An expression like Eq. 13.1 is called a *Lennard-Jones potential.* In most cases, this is not appropriate for the interaction of a chemisorbed atom with the atoms of a solid, but I use it because it has correct qualitative features. I am not trying to perform an accurate calculation, but to develop and explain rate theory. If I were serious about this calculation, I would use a numerical method, from quantum chemistry, to calculate the potential energy surface.

The Lennard-Jones potential contains two parameters, the energy ϵ and the distance σ. Reasonable values for them are $\epsilon = 0.3$ eV and $\sigma = 2.8$ Å. I show in Fig. 13.2 what this potential looks like. The plot was made in Workbook SM13.1. The shape of the potential is quite general: other potentials look similar.

What is this shape telling me? The potential has a minimum at an interatomic distance d_0. This means that, when put in contact, the two atoms will make a diatomic molecule with the bond length d_0. For a Lennard-Jones potential, $d_0 = \sqrt[6]{2}\,\sigma$ (see Exercise 13.1).

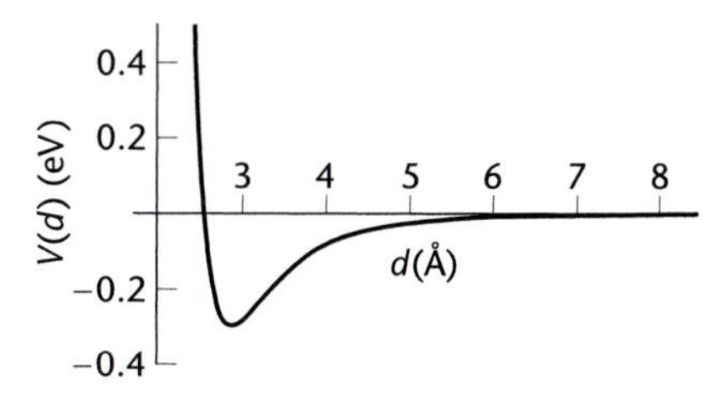

Figure 13.2 The Lennard-Jones potential energy of interaction between two atoms. I used Eq. 13.1 and the parameters $\epsilon = 0.3$ eV and $\sigma = 2.8$ Å.

Exercise 13.1

Find a formula giving the value of the bond length for two atoms interacting through a Lennard-Jones potential. Calculate the numerical value of the bond length in cm and in Å for $\epsilon = 0.3$ eV and $\sigma = 2.8$ Å. See Workbook SM13.2 and Workbook SM13.3.

When the distance between the two atoms becomes very large the interaction energy between the atoms becomes zero. For the parameter values used here, this happens when the interatomic distance is larger than 6 Å, as you can see in Fig. 13.2.

Recall that adding a constant to potential energy does not change the value of any physical observable. This is so because the force of the interaction between the atoms is given by

$$F = -\frac{\partial V(d)}{\partial d}$$

Since the force determines the behavior of the particles, no change in force means that I cannot tell the difference between the motion due to $V(d)$ and the motion due to $V(d) + C$. In the Lennard-Jones potential this constant was chosen so that the potential energy is *zero when the atoms no longer interact.*

The binding energy E_b of the molecule is

$$E_b = V(d_0) - V(\infty) \tag{13.3}$$

Here $V(d_0)$ is the energy of the molecule and $V(\infty)$ is the energy of the separated atoms (which is zero). For the Lennard-Jones potential, $E_b = -\epsilon$ (see Exercise 13.2).

Exercise 13.2

Find a formula giving the value of the binding energy of a diatomic molecule, as a function of ϵ, for two atoms interacting through the Lennard-Jones potential. Calculate the numerical value of the bond energy in eV and erg. See Workbook SM13.2 and Workbook SM13.3.

Imagine now that you can grab the two atoms in the molecule and push them closer or pull them away from each other. In other words you try to shorten or lengthen the bond length. Since the bond length corresponds to a minimum in the potential energy, either action will increase the potential energy of the molecule. To cause this increase you will have to provide work, which means that you have to exert a force; the molecule opposes an attempt to either increase or decrease its bond length. This potential describes a love–hate relationship: the partners say to each other, don't get too close and don't go too far; stay at the distance at which the energy is minimum.

Exercise 13.3

Obtain the same conclusions by analyzing the force between the particles $F = -\partial V/\partial d$. Don't calculate the force. You only need its sign; get it from the slope of the graph.

Exercise 13.4

Make a plot of the force between the atoms in a diatomic molecule described by the Lennard-Jones potential. Use for σ and ϵ the values used to make Fig. 13.2.

Exercise 13.5

A 300-pound man steps on a diatomic molecule. The molecule is upright on a hard floor and the fellow has very hard soles on his shoes; neither the floor nor the soles deform. How much is the distance between the atoms changed? Note that the bond distance of the compressed molecule is such that the weight of the man is balanced by the repulsive force between the molecules. Be careful with the units. The force exerted by the weight of the man is mg where m is the mass and g is the acceleration of gravity.

Exercise 13.6

You tie an atom of the diatomic molecule to the ceiling and you hang from the other atom a 300-pound weight. How far will the molecule stretch? The force of gravity (mg, where g is the acceleration of gravity and m is the mass of the weight) will cause an increase of the bond length while the force binding the atoms (whose value is $-\partial V(r)/\partial r$) opposes this stretch. Equilibrium is reached if the absolute values of the two forces are equal.

§9. *The Energy of Interaction of the Adsorbed Atom with the Solid.* The energy of interaction between the chemisorbed atom and all the atoms of the solid is

$$V(\mathbf{r}) = \sum_{i=1}^{n} v(d_i) \tag{13.4}$$

Here $v(d_i)$ is the Lennard-Jones potential (see Eq. 13.1), d_i is the distance from the adsorbed atom to the atom i in the solid (see Eq. 13.2), and the sum is over all the atoms in the solid. Since the Lennard-Jones potential goes to zero at large distances, I only need to sum over those atoms of the solid that are close to the adsorbed atom.

§10. *The Structure of the Solid.* To calculate $V(\mathbf{r})$, I need to know the positions $\mathbf{r}_i$ of the atoms in the solid. These are defined in terms of the quantities a and z_0 explained in Fig. 13.3. I take $z_0 = a/2$ and $a = 2.8$ Å. The value of a is the same as that of σ in Eq. 13.1, by accident. The positions of the atoms in the solid, corresponding to these values of a and z_0, are calculated in Workbook SM13.4. Pictures of the atomic positions, in the first two layers of the solid, are made in Workbook SM13.5.

§11. *The Potential Energy* $V(\mathbf{r})$. The positions of the atoms in the solid, together with the Lennard-Jones potential, completely determine $V(\mathbf{r})$, which is calculated in Workbook SM13.6. To generate a formula for $V(r)$, I calculate the coordinates of the atoms in the solid (see Fig. 13.4 and Workbook SM13.4) and then I perform the sum indicated by Eq. 13.4.

The qualitative features of $V(\mathbf{r})$ produced by this calculation are correct, even though the model is not accurate. The inaccuracy of the model does not affect the discussion that follows.

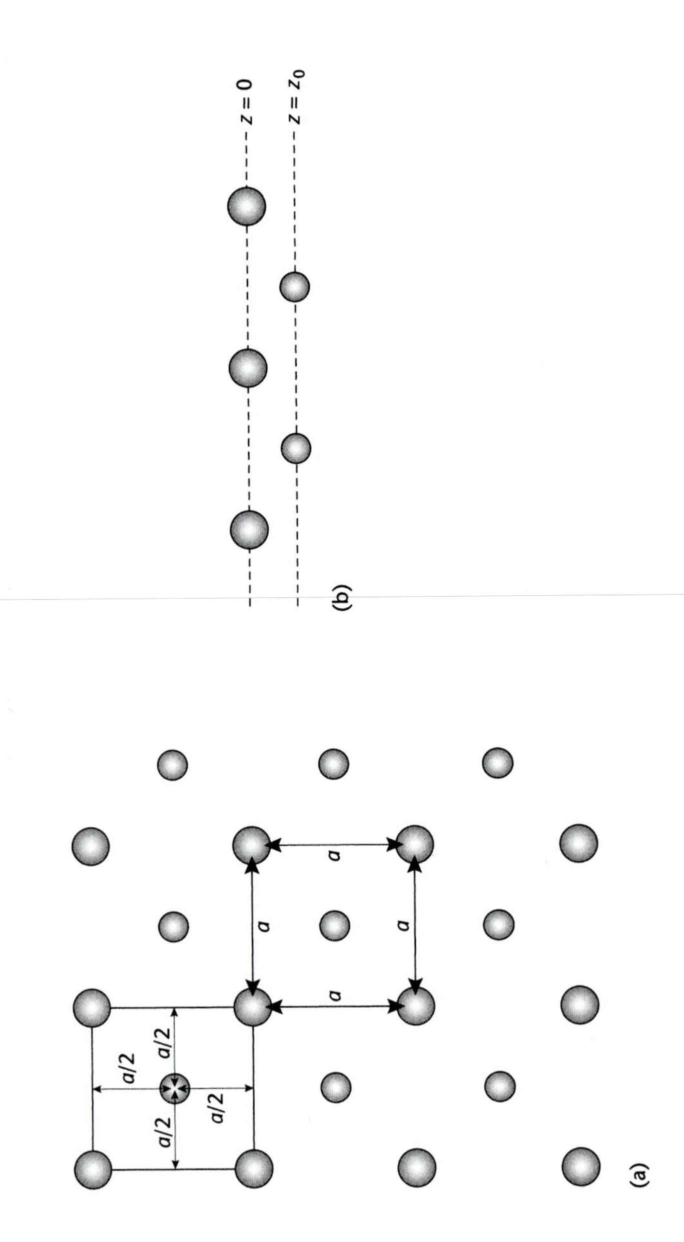

Figure 13.3 (a) The top two atomic layers at the surface of the solid, seen when looking down at the surface. The larger atoms are in the top layer. (b) The atoms in the top two layers, seen from the side. Not all surfaces of a given solid have this structure, and the structure of most solids is more complicated.

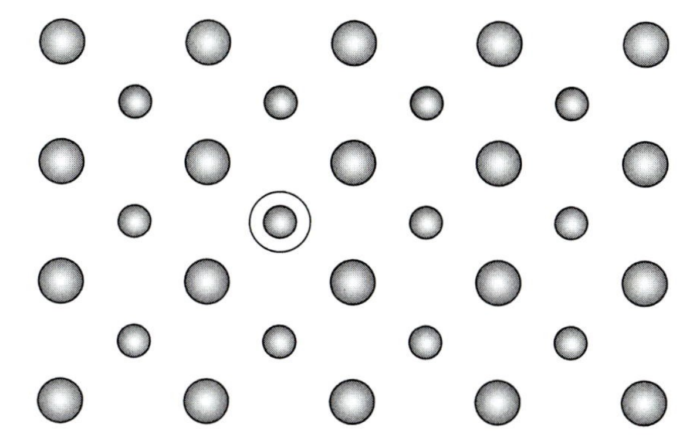

Figure 13.4 The solid atoms included in the calculation of $V(\mathbf{r})$ with Eq. 13.4. The spheres representing the atoms in the top layer are larger than those representing the atoms in the second layer. The equilibrium position of the adsorbed atom is shown by the circle.

§12. *The Shape of* $V(\mathbf{r})$. The shape of $V(\mathbf{r})$ is determined by competition between the atoms in the solid for the affection of the chemisorbed atom. Each atom in the solid wants to form a bond of the appropriate length with the chemisorbed atom. Obviously, it is impossible for the chemisorbed atom to satisfy so many conflicting desires. The atom does what most politicians do: it settles in a compromise position that minimizes *the sum* $V(\mathbf{r})$, not each individual pairwise interaction $v(d)$. It tries to satisfy all atoms, by not giving full satisfaction to any individual atom. However, as in politics, some atoms are more influential than others. Since the interaction between atoms fades with distance, the atoms closest to the chemisorbed atom have the largest influence on its position.

Exercise 13.7

Calculate $V(\mathbf{r})$ under the assumption that only the five solid atoms closest to the adsorbed atom contribute to the potential energy. Are you making a large error by doing this? Determine which atoms are less influential.

Since I know $V(\mathbf{r})$, I can calculate the position $\mathbf{r}$ for which $V(\mathbf{r})$ has a minimum. This calculation is done in Workbook SM13.7 and gives $\mathbf{r}_A = \{x = 4.203, y = 4.200, z = 2.084\}$ Å. $\mathbf{r}_A$ is the position where the atom sits still if the temperature is 0° K. If the temperature is not too high, the adsorbed atom oscillates around $\mathbf{r}_A$.

The calculation shows that the adsorbed atom prefers to sit in a four-fold site (at the center of a square formed by four atoms in the top layer). If the solid surface is large, then $V(\mathbf{r})$ has a minimum at each four-fold site. At the edge of the surface the minimum is no longer in the center of the square.

§13. *The Symmetry of the Surface and the Position of the Chemisorbed Atom (a Side Remark).* One might be tempted to think that symmetry will force the chemisorbed atom to take a fourfold position. The chemisorbed atom located somewhere in the unit cell interacts most strongly with the four atoms bordering the cell. Since these four atoms are identical, the distances between the chemisorbed atom and these four atoms should be equal. This requirement places the atom in the four-fold site, which preserves the four-fold symmetry of the system. This position is also favored by the requirement that the potential energy should be minimized. The atom binds to five solid atoms (the four forming the square and the one in the second layer), and each bond lowers the potential energy.

However, this argument is not correct. Imagine that the bond between the chemisorbed atom and the solid atoms is very strong, and the bond length between them is very short. Moreover, the distance between the solid atoms is larger than the bond length between the chemisorbed atom and a solid atom. In this case I expect that the chemisorbed atom will not sit in the four-fold site. It might prefer to be closer to one of the surface atoms (see the picture in the next exercise). What happens then to the square symmetry of the system? It is still preserved. There are four equivalent binding sites in each unit cell, and these binding sites form a square.

Exercise 13.8

I suspect that the adsorbed atom does not always sit in the middle of a square. If it binds strongly to the atoms in the solid, and the distance a between the solid atoms is much larger than σ, the atoms could sit at one of the four locations shown below.

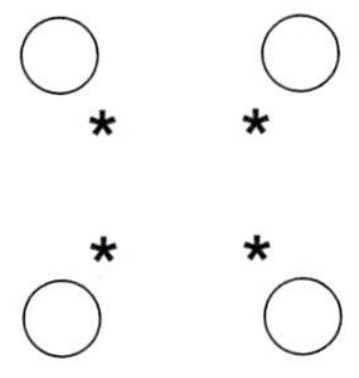

Note that these locations preserve the symmetry of the lattice: they form a square. Choose a large value for ϵ ($\sim$ 1 eV) and $\sigma \ll a$, and calculate the position of the

atom to check whether my guess is right. Experiment with several values of ϵ and σ.
Hint. To find the position of the chemisorbed atom minimize $V(\mathbf{r})$.

§14. *The Topology of the Surface Around the Binding Site.* To develop the theory of the rate constant we need to get some information about the topology (the shape) of the potential-energy surface $V(\mathbf{r})$ around the binding site. The easiest way to study the shape of the surface $V(x,y,z)$ is to make a plot and look at it. Unfortunately, $V(x,y,z)$ is a function of three variables, and we only know how to plot functions of two variables.

We can get a pretty good idea of the shape of the surface around the binding site by plotting cuts through this surface. To explain what I mean by this, I denote by $\{x_A, y_A, z_A\}$ the coordinates of the binding site (the four-fold site), which are equal to $\{x_A = 4.203, y_A = 4.200, z_A = 2.084\}$ Å. The coordinate z is perpendicular to the surface and x and y are parallel to it. I can ask, for example, how the potential energy changes if I change z and keep $x = x_A$ and $y = y_A$. I can see this by plotting $V(x_A, y_A, z)$ as a function of z. The plot will tell me what happens to the potential energy when I pull the chemisorbed atom away from the surface or I push it towards it.

You can see a plot of $V(x_A, y_A, z)$ as a function of z in Fig. 13.5. The plot was made in Workbook SM13.8. The picture contains no surprises: it is similar to that of the interaction between two atoms (see Fig. 13.2). We expect the energy to increase if we push the atom into the surface or if we pull it away from it. If we pull the atom

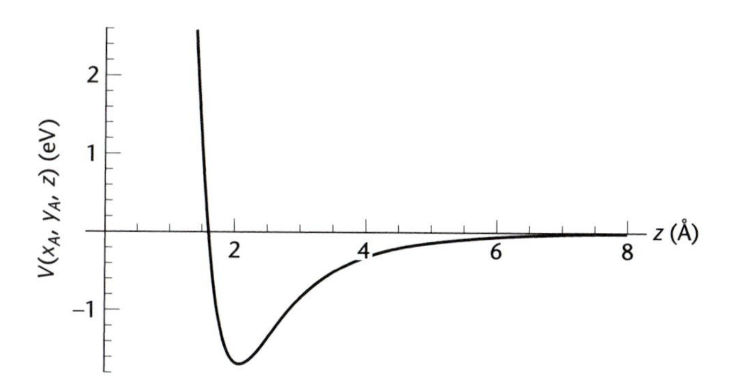

Figure 13.5 The potential energy surface as a function of the distance between the chemisorbed atom and the solid surface.

too far from the surface, the bond between the atom and the surface breaks, and the potential energy becomes zero.

While the shape of $V(x_A, y_A, z)$ is similar to that between the chemisorbed atom and the atoms of the surface, the numerical value of the binding energy and the bond length are not. The distance z at which $V(x_A, y_A, z)$ has a minimum is $z = 2.0827$ Å. The binding energy of the chemisorbed atom to the surface is $V(x_A, y_A, z_A) - V(x_A, y_A, z = \infty) = V(x_A, y_A, z_A) = -1.679$ eV. These values have been calculated in Workbook SM13.7.

The binding energy of the chemisorbed atom to the surface is much larger than the binding energy of the chemisorbed atom to one atom of the solid.

Exercise 13.9

The binding energy of the chemisorbed atom to the solid is $\varepsilon_a = -1.679$ eV. The parameter ϵ in the Lennard-Jones potential used to calculate ε_a is 0.3 eV. Can you assume, as a crude approximation, that the chemisorbed atoms binds to the nearest five atoms of the solid? Why does this assumption fail?

Exercise 13.10

Plot the potential energy surface as a function of z, when $x = x_A + 0.3$ Å, $y = y_A + 0.4$ Å. Remember that $x_A = 4.203$ Å and $y_A = 4.200$ Å. Try to answer, before performing the calculations, the following questions: (a) Is the shape of the plot the same as the one shown in Fig. 13.5? (b) Is the absolute value of the minimum energy in the plot smaller or larger than $V(x_A, y_A, z_A)$? In other words, does it take less or more energy to remove the atom from the surface in the case when x and y differ from x_A and/or y_A?

§15. *A Slight Detour: Adsorption and Desorption.* I will make a detour to discuss another "chemical reaction" in which a chemisorbed atom can participate. Consider the following experiment. At low temperature, a gas (say, carbon monoxide) is put in contact with a clean solid surface (say, nickel). I can, for example, drop some Ni powder in the gas. As soon as contact between the powder and the gas is made, the pressure of the gas starts decreasing and ultimately settles to a lower value than the initial pressure. The decrease in pressure tells me that some CO molecules disappear from the gas (remember $p = Nk_BT/V$, where p is the pressure, V is the volume, T is the temperature, N is the number of molecules in the gas, and k_B is the Boltzmann constant). Where did the missing molecules go? They are stuck to the Ni surface. If you want to sound scientific you say that some of the CO molecules originally in the gas have been *adsorbed* by the nickel surface. The process is called *adsorption*.

To convince myself that the molecules are adsorbed on the surface, I perform the following experiment. I cool the surface to a very low temperature (e.g. 70 K) and pump the CO gas out from the apparatus, until the pressure is very low (pressures of 10^{-11} Torr are reached in a well-equipped laboratory). I am inclined to think that if the pressure is so low, I pumped all the CO molecules out of the system. Not so! If I heat the nickel surface, I find that as the surface temperature increases, CO molecules appear in the gas phase. I detect their presence by the increase of pressure in the system. This process of removal of adsorbed molecule from a solid surface is called *desorption*.

Adsorption takes place because the energy of a CO molecule bound to the nickel surface is lower than the energy of the same molecule in the gas. The energy required to remove an adsorbed molecule from the surface is called the *adsorption energy*. If this energy is large (larger than about 0.2 eV), we say that the molecule is *chemisorbed*. If this energy is small, we say that the molecule is *physisorbed*. Noble gas atoms, which are unable to make strong chemical bonds, are physisorbed, no matter what the solid is made of. Since gold is rarely willing to make chemical bonds, most molecules are physisorbed on a Au surface. Chemisorption is more common than physisorption, since most metal surfaces make chemical bonds with most molecules.

Adsorption is often used in industry. The most important application is in heterogeneous catalysis, where a metal or an oxide surface is used to increase the rate of a chemical reaction. Some of the most widely used chemicals, such as gasoline, ammonia, sulfuric acid, methanol, and ethylene oxide, are made by catalysis. Catalysis takes place only if the reactants adsorb on the metal surface and the products desorb from it. Getting favorable adsorption and desorption properties is part of the design of a good catalyst.

Adsorption is also used to remove unwanted gases from an enclosure. Porous carbon or silica that have enormous surfaces (hundreds of square meters per gram of material) can be prepared. The carbon adsorbs organic substances from water, purifying it. Porous silica is used to adsorb water from industrial gases, in situations where water is unwanted (you don't want water in a pipe transporting methane in the winter; it would form ice and clog the pipe).

Why am I concerned with desorption here? Because I wanted to point out that we have calculated the adsorption energy of our chemisorbed atom. Indeed, the adsorption energy is $E_{ads} = V(x_A, y_A, z_A) - V(x_A, y_A, \infty)$. This is the energy of the system when the atom is absorbed on the surface in its favorite position $\mathbf{r}_A = \{x_A, y_A, z_A\}$, minus the energy when the atom is removed from the surface (its z-coordinate is infinity). Because $V(x_A, y_A, \infty) = 0$, the chemisorption (or absorption) energy is $V(x_A, y_A, z_A)$, which we have already calculated to be -1.679 eV.

I must emphasize that this is the adsorption energy for the simplified model used here, where the solid atoms do not change their position when the chemisorbed atom is removed. This is a fairly crude approximation. When I remove the chemisorbed atom the solid's atoms to which the chemisorbed one was bound, will shift their position. This shift will produce a change ΔE in the potential energy. Because of this change the adsorption energy will be lower than -1.679 eV, by ΔE.

§16. *The Shape of the Potential Energy Surface Around the Fourfold Binding Site.* Let us return to our chemisorbed atom and its travels along the surface. I know now what happens to the potential energy when I remove the chemisorbed atom from the surface and keep the coordinates x and y constant. To add to my understanding of the shape of the potential energy surface around the four-fold position, I plot the function $V(x, y, z_A)$. Now I let the chemisorbed atom move parallel to the surface, while keeping the surface-atom distance z constant (and equal to the equilibrium distance z_A).

Because z is held constant the potential energy depends only on the variables x and y. There are two ways of plotting a function of two variables. One is a contour plot, in which I display lines joining points having the same potential energy. Those of you who hike may have used a topographic map, which displays contours of equal height. The contour plot of $V(x, y, z_A)$ (made in Workbook SM13.8) is shown in Fig. 13.6.

As I expect, the contour map shows that the surface has a minimum at the four-fold site, marked in the graph by A. To reach A from any point on the surface, I have to go down on the potential energy hill. The transition state point TS is a different matter. A point moving away from TS in the x direction (towards A) goes downhill. Therefore TS is a maximum of the potential energy surface, with respect to x. A point moving away from TS, in the y direction, towards the atom in the upper right corner of the figure, goes uphill. Therefore, TS is a minimum, on the potential energy surface, with respect to the variable y. I will study shortly, and in more detail, the shape of the potential energy surface around the point TS.

For those of you who don't use topographic maps, I draw a different plot in Fig. 13.7. In this plot the point z represents the value of the function $V(x, y, z_A)$ for any pair of values of $\{x, y\}$.

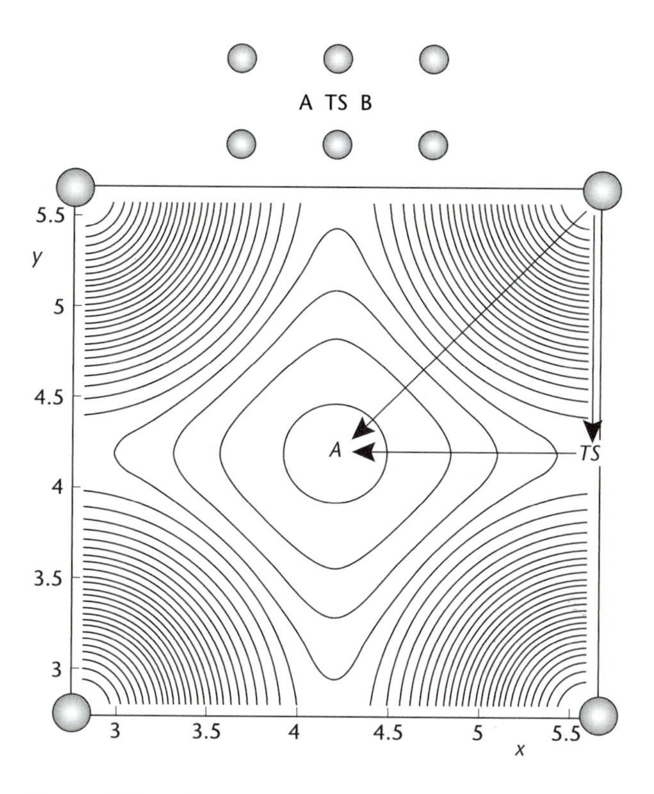

Figure 13.6 The potential energy surface $V(x, y, z_A)$ when the chemisorbed atom moves parallel to the surface at a constant height $z_A = 2.0827$ Å. The inset above the contour plot shows a few of the surface atoms, to display the unit cell A (enclosed by the dotted square) and the position of the transition state. The contour map is drawn for x and y varying over the cell A. The arrows show the "downhill" directions on the surface, in one quadrant of the picture. The other quadrants are similar because of symmetry.

You can see in this figure that the surface has a minimum in the four-fold site (which is in the center of the figure) and that the TS point (shown in the figure by an empty circle) is a saddle point.

Note that the shape of the potential energy surface around the four-fold site is that of a bowl. No matter in which direction the point $\{x, y\}$ moves away from the four-fold site, the potential energy surface is *a parabola* centered at the four-fold site. This is true as long as the atom does not travel far from the minimum. We will

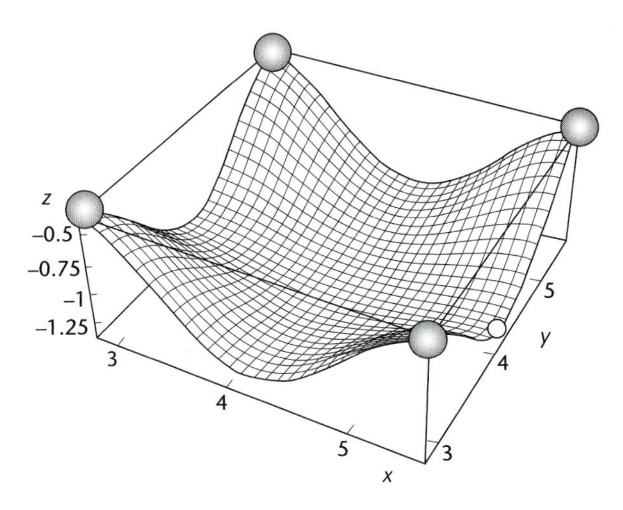

Figure 13.7 The potential energy surface $V(x, y, z_A)$ when the chemisorbed atom moves parallel to the surface at a constant height $z_A = 2.0827$ Å. The coordinate z in the plot shows the values of the function $V(x, y, z_A)$ for each pair of points $\{x, y\}$. This is the surface shown by the contour plot in Fig. 13.6.

make use of this property later, when we calculate the partition function of the atom in cell A.

Exercise 13.11

Plot $V(x = 4.203, y, z = 2.084)$ as a function of y, and $V(x, y = 4.2, z = 2.084)$ as a function of x. Check whether the two plots resemble parabolas for values of x or y around 4.202 Å.

§17. *Find the Coordinates of the Transition State.* I will first make contact with and review some of the things you learned in the previous chapter. When the chemisorbed atom is in the cell A (indicated by the square in Fig. 13.1), we say that the atom and the solid together form the molecule $\mathcal{A}$. When the atom is in the cell B (in Fig. 13.1), we say that the system forms molecule $\mathcal{B}$. The migration of the atom from cell A to cell B is an isomerization reaction. In the previous chapter I argued that any reactive trajectory going from A to B has to pass over a ridge on the potential energy surface. The point on this ridge having the lowest potential energy is the transition state. I denote here the coordinates of the transition state by $\{x_{TS}, y_{TS}, z_{TS}\}$.

Now I want to find the transition state (i.e. the numerical values of $\{x_{TS}, y_{TS}, z_{TS}\}$) for the chemisorbed atom migrating from cell A to cell B.

Finding the transition state on a general potential energy surface is a delicate matter and requires quite a bit of sophistication. The system I study here was intentionally picked to make finding of the transition state easy. By symmetry, I expect the transition state to be located in the middle of the line separating the cell A from cell B (in Fig. 13.1 this line joins atom a to atom b). In what follows I will do a few calculations to verify that this is the case for the potential energy surface $V(x, y, z)$ of this system.

The line joining the atoms a and b (please do not confuse the surface atom a with the lattice constant a) in Fig. 13.1 passes through $x = 2a = 5.6$ Å (remember that the lattice constant $a = 2.8$ Å). The middle of the line between the atoms a and b in Fig. 13.1 has the coordinate $y = a + \frac{a}{2} = 4.2$ Å. This gives me two of the coordinates of the transition point $x_{TS} = 5.6$ Å and $y_{TS} = 4.2$ Å. To find the third, I use the fact that the energy of the system at the transition state has a minimum with respect to the z coordinate. This means that we can find the value of z_{TS} by finding the minimum of the function $V(x_{TS}, y_{TS}, z)$ with respect to z. I performed this calculation in Workbook SM13.9. The result is $z_{TS} = 2.46109$ Å.

In conclusion, the position of the transition state is $\{x_{TS} = 5.5 \text{ Å}, y_{TS} = 4.2 \text{ Å}, z_{TS} = 2.46109 \text{ Å}\}$. The potential energy when the atom is at the transition state is $E_{TS} = V(x_{TS}, y_{TS}, z_{TS}) = -1.224$ eV (see Workbook SM13.9). As I suggested in the previous chapter, the difference between the energy E_{TS} and the energy E_A (the binding energy of the atom to the four-fold site on the surface) is very close to the activation energy E_a. I have already argued, in the previous chapter, that this quantity will play an important role in rate theory. Its value is calculated in Workbook SM13.9 and is

$$E_{TS} - E_A = -1.224 \text{ eV} - (-1.679) \text{ eV} = 0.455 \text{ eV} \tag{13.5}$$

§18. *Check Whether this Guess of the Position of the Transition State is Correct.* If I have guessed the position of the transition state correctly, then this point ought to be a maximum of the potential energy with respect to x and a minimum with respect to z and y. I determined z_{TS} to give the minimum of $V(x_{TS}, y_{TS}, z)$, so the condition on z is satisfied. I want to convince myself that $V(x, y_{TS}, z_{TS})$ has a maximum with respect to x and that $V(x_{TS}, y, z_{TS})$ has a minimum with respect

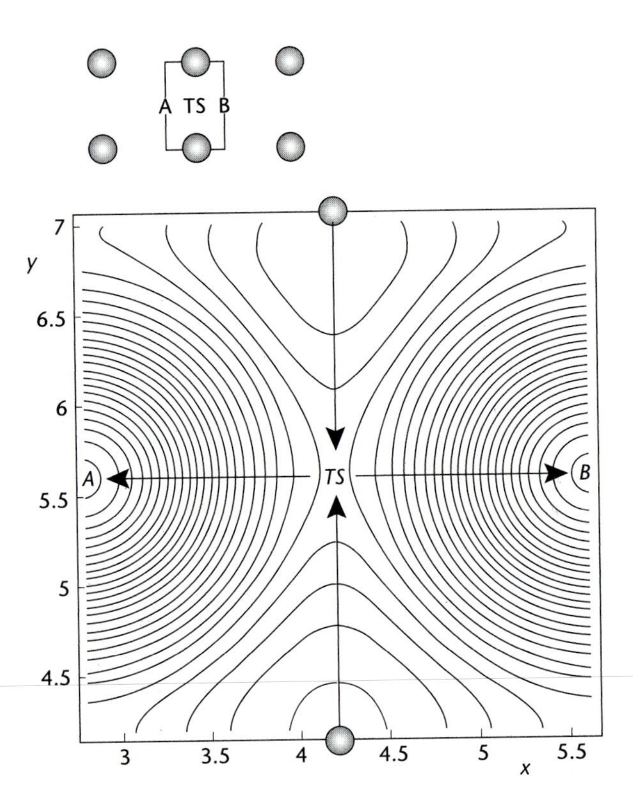

Figure 13.8 A contour plot of the potential energy surface $V(x, y, z_{TS})$ when the chemisorbed atom moves parallel to the surface, at a constant height $z_{TS} = 2.461$ Å. The range of variation of x and y is shown by the square in the inset above the contour plot. The arrows show the downhill direction around the transition state. Note that TS is a maximum of $V(x, y_{TS}, z_{TS})$ in x and a minimum of $V(x_{TS}, y, z_{TS})$ in y. The surface around TS has a saddle shape (the y-direction is along the spine of the horse).

to y. I have checked that this is true, by making plots of these functions, in Workbook SM13.9. These calculations convince me that my guess of the position of the transition state is correct.

§19. *The Shape of the Potential Energy Surface Around the Transition State.* To understand the shape of the potential energy of the surface I show first, in Fig. 13.8, a contour plot of $V(x, y, z_{TS})$. Those of you who are not used to contour plots can take a look at Fig. 13.9. This shows a picture of the surface $V(x, y, z_{TS})$

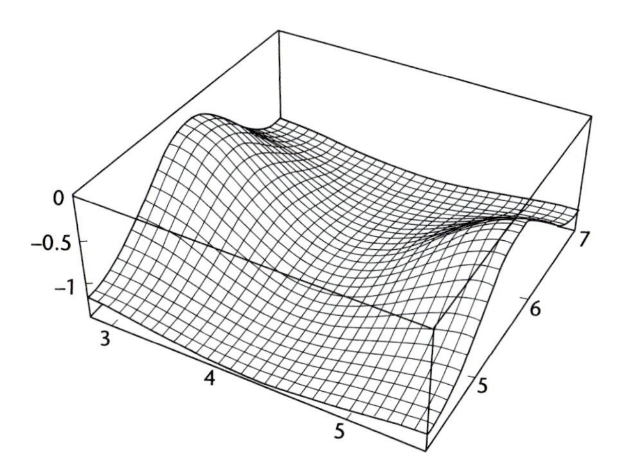

Figure 13.9 The potential energy surface $V(x, y, z_{TS})$ when the chemisorbed atom moves parallel to the surface at a constant height $z_A = 2.0827$ Å. The coordinate z in the plot shows the values of the function $V(x, y, z_A)$ for each pair of points $\{x, y\}$. This is the surface shown by the contour plot in Fig. 13.8.

(the same as the one shown in the contour plot). In this picture you can see clearly the saddle point in the middle of the figure where the TS point is located.

§20. *Summary.* In this chapter we have accumulated all the knowledge needed for the chapters that follow. We have constructed a potential-energy surface for the moving atom and have studied its shape around the binding site and the transition state. Understanding this is essential for understanding rate theory. The way we have gathered this information gives an indication of how this kind of problem is solved in practice.

14

TRANSITION STATE THEORY: THE RATE CONSTANT

Introduction

§1. In Chapter 12, I presented a qualitative theory of a chemical reaction. The key quantity is the potential energy surface, which is a function of the $3N$ coordinates of all the atoms in the system (N is the number of atoms). A molecular configuration is represented by a point $C = \{\mathbf{r}_1, \mathbf{r}_2, \ldots, \mathbf{r}_n\}$ in a $3N$-dimensional space. I have argued that the structure of the reactants and the products are represented by minima in the potential-energy surface of the system. These minima are separated by an energy ridge. A reaction takes place when the point representing the positions of the atoms travels from the minimum corresponding to the reactants to the minimum corresponding to the products. A trajectory that starts in the reactant region and ends in the product one is called a *reactive trajectory*.

A reactive trajectory must, by definition, cross the ridge separating the two minima. Since the system is in equilibrium throughout the reaction, the probability that the point representing the system reaches the ridge can be calculated from equilibrium statistical mechanics. In the previous chapters, I have suggested that the rate law and the rate constant can be calculated by a theory that uses the probability that the representative point reaches the ridge, and the velocity with which the ridge is crossed. In this chapter I will show how such a theory is constructed.

The general presentation of the theory is hampered by tedious mathematics. Because of this, I decided that it is better to derive the theory for the example of an atom moving from site-to-site, on a solid surface. In this example, the mathematical complications are reduced to a minimum. All you need to know about this example has been presented in Chapter 13. In particular, we have constructed a potential-energy surface, and we have located the minima, the ridge, and the transition state on it.

The example illustrates the principles involved in the theory; the generalization to more complex cases is fairly obvious (though mathematically tedious). Whenever it is warranted, I will point out to you the kind of complications that appear in more general situations.

§2. *Site-to-Site Hopping as a Problem in Kinetics.* I have suggested several times that the problem of site-to-site hopping by a chemisorbed atom is a problem in chemical kinetics. It is now time to make this connection more precise.

Imagine that I have deposited N atoms on a perfectly flat solid surface, at a time that I will call zero. I make sure that the concentration of these atoms on the surface is low, so that they do not interact with each other (a concentration of 10^{11} atoms per cm^2 will achieve this). If two atoms happened to be, by accident, too close to each other, they will be ignored.

The atoms are in equilibrium with the solid. Since the temperature is kept constant and I do not act on the atoms with any outside force, the equilibrium is maintained throughout the experiment.

I will call the cell in which each atom is initially located, a cell (or a site) of type A (Fig 14.1). As explained in Chapter 13 these atoms will rattle around for a while and from time to time will gather enough energy to leave cell A and settle in a neighboring cell. I will call this neighboring cell, a cell of type B. Any one of the four cells surrounding cell A is a cell of type B. Because of the symmetry, the rate of jumping from A to a neighboring cell is the same for all four B cells.

Because the atoms will sooner or later jump out of cell A, the number of atoms in cell A is a function of time $N_A(t)$. Since each jump creates an atom in cell B, the number of atoms in cell B is also a function of time $N_B(t)$. At time zero I have $N_A(0) = N$ and $N_B(0) = 0$.

I have explained several times that a jump from a cell of type A to one of type B is an isomerization reaction. Phenomenological chemical kinetics tells me that the

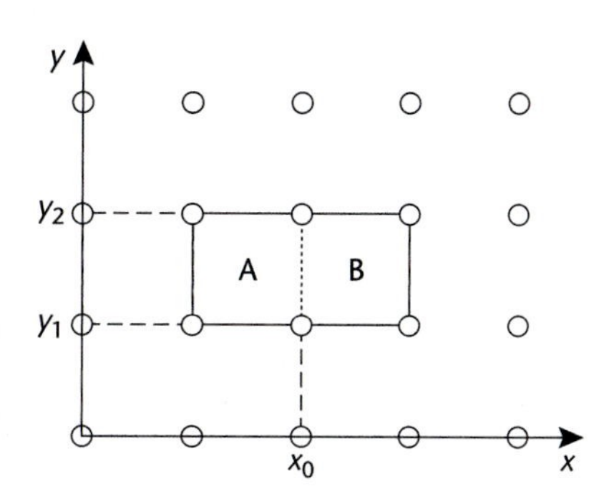

Figure 14.1 The dotted line shows the position of the potential energy ridge between A and B. The solid lines delimit sites A and B. It is assumed that the atom cannot cross the solid lines.

rate of change of N_A can be written as

$$\frac{dN_A}{dt} = \left(\frac{dN_A}{dt}\right)_- + \left(\frac{dN_A}{dt}\right)_+ \tag{14.1}$$

$(dN_A/dt)_-$ is the number of atoms that *jump out* of A per unit time. Such jumps decrease the number of atoms in A and therefore $(d\,N_A/dt)_-$ is negative. $(d\,N_A/dt)_+$ is the number of atoms *jumping into* A from B in unit time. This is a positive number, because such jumps increase N_A.

To simplify the problem note that once an atom leaves A, and settles in B, we pretend that it no longer exists. Its return to A, if it were to happen, is another "chemical reaction" whose rate should be treated separately. Because of this we can set

$$\left(\frac{dN_A}{dt}\right)_+ = 0$$

We are examining only the rate of leaving cell A in an one-way trip; the return is another trip.

§3. *The Rate Equation.* Phenomenological chemical kinetics tells us that the jumping rate is proportional to the population N_A of jumpers:

$$\left(\frac{dN_A}{dt}\right)_- = -k_{A\to B}N_A \tag{14.2}$$

If more atoms are present in the site A, it is more likely that we will observe one jump, and the jumping rate will be higher. Eq. 14.2 was found by performing and analyzing experiments.

The constant $k_{A \to B}$ is the rate constant for the forward reaction A→B. In this chapter I develop a theory that derives Eq. 14.2 and provides a recipe for calculating $k_{A \to B}$. It will turn out that this theory is also able to calculate the rate constant $k_{B \to A}$ that appears in the rate equation

$$\left(\frac{dN_B}{dt}\right)_- = -k_{B \to A} N_A \tag{14.3}$$

The Derivation of the Rate Equation and of a Formula for the Rate Constant

§4. *The Dividing Surface.* I will now concentrate on $(dN_A/dt)_-$, which is the rate by which the molecules *leave* site A. To leave site A, the atom must cross the ridge separating site A from site B. Let us study this process.

In the system we consider here, the ridge is on the dotted line in Fig. 14.1, located between cells A and B. Now place a plane, as in Fig. 14.2, that is perpendicular to the surface of the solid and passes through the ridge. A reactive trajectory (any trajectory going from A to B) must pass through this plane. We call this plane the *dividing surface.*

It is easy to define this plane mathematically. An atom at $\{x, y, z\}$ is on the dividing surface if

$$x = x_0$$

$$y \in [y_1, y_2] \quad (y \text{ is between } y_1 \text{ and } y_2)$$

$$z \text{ can have any value}$$

where x_0, y_1, and y_2 are defined in Fig. 14.1.

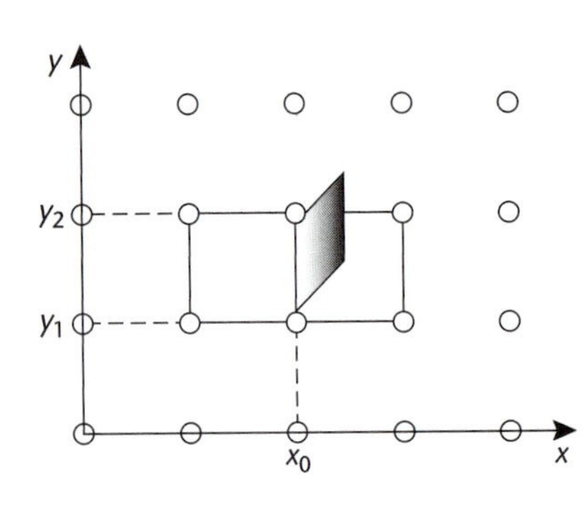

Figure 14.2 The dividing surface separating the reactant (A) from the product (B). All reactive trajectories must pass through this surface. The dividing surface is a plane perpendicular to the surface of the solid.

I must emphasize that the ridge is a straight line and the dividing surface is a plane only because I am working with a very simple example, which has a high symmetry. A more complicated example (but still very simple) is examined in Supplement 14.1.

§5. *The Activated Complex.* As we have discussed, the ridge is like a continental divide, separating the positions for which the atom is in site A (reactant) from those for which it is in site B (product). If the atom is on the A side of the dividing surface, the forces acting on it will push it towards the center of cell A (like a ball rolling down the hill). If the atom is on the B side of the dividing surface, the interaction with the solid will push it towards the center of cell B. The reaction takes place when the atom passes through the dividing surface.

Clearly, the probability that the atom reaches the dividing surface ought to be an important element in rate theory. In our particular case, this means the probability that $x = x_0$ and $y \in [y_1, y_2]$.

It is, however, meaningless to ask for the probability that x is exactly equal to x_0. That probability is zero! A meaningful question is: what is the probability that

$$x \in \left[x_0 - \frac{\delta}{2}, x_0 + \frac{\delta}{2}\right]$$

where δ is a very small number?

Geometrically this means that I want to know the probability that the atom is between the two planes passing through $x_0 - \delta/2$ and $x_0 + \delta/2$ (see Fig. 14.3). When the atom is located between these planes, we say that it forms a *transition complex* A^* or an *activated complex*. The system forms a transition complex when the coordinates $\mathbf{r} = \{x, y, z\}$ of the atom satisfy the condition

$$x_0 - \tfrac{\delta}{2} \leq x \leq x_0 + \tfrac{\delta}{2}$$

$$y \in [y_1, y_2]$$

z can have any value

This is the same as saying that the atom is in the slab defined by the two planes.

§6. *The Number of Atoms Forming an Activated Complex.* To study the statistical properties of the system, I use an ensemble. I imagine that I have a huge number N of identical systems, each having the same temperature T and each consisting of a surface with one atom in cell A. The system is in

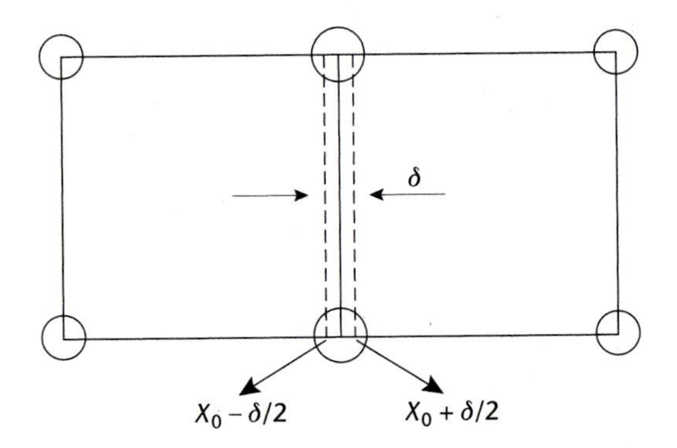

Figure 14.3 The dashed lines indicate where the two planes defining the transition complex intersect the solid surface. The planes are perpendicular to this surface (and parallel to the dividing surface).

equilibrium (we do not act on it, and it has been sitting for a sufficiently long time in a thermostat). In some systems in the ensemble, the atom is in cell A, and in others it is in the strip A* surrounding the dividing surface. There are N_A atoms is in cell A and N^* in the transition state region.

Because the system is in equilibrium, these numbers are related (see Chapter 11). We have

$$\frac{N^*}{N_A} = \frac{q^*}{q_A} \tag{14.4}$$

where q^* and q_A are the partition functions of the atom in the strip A* and the cell A, respectively. We have methods for calculating q^* and q_A, so — at least in principle — these two quantities are known. Therefore, I can calculate N^* from Eq. 14.4:

$$N^* = \frac{q^*}{q_A} N_A \tag{14.5}$$

This an important relation and I will use it later.

Because an atom in the strip A* has a substantially higher potential energy than an atom in cell A (or cell B), I expect N^* to be much smaller than N_A. The atoms have a hard time climbing the potential-energy hill to get into the strip A*. This is not an unfamiliar situation. At any given time, there are very few people on mountain peaks over 14,000 feet high. Lifting your body to that height increases your potential energy in the gravitational field ($V = \text{mass} \times g \times \text{height}$). Like atoms, most people are unlikely to perform the work needed for lifting themselves above 14,000 feet. Only a few have the willpower and stamina to do it. In defense of people, I should mention that they have excuses that atoms lack: poor health, age, short vacations, or stopping in Las Vegas and never reaching the Sierra Nevada.

Many textbooks will tell you that Eq. 14.5 is the major approximation in this theory. That is not true. Eq. 14.5 is exact, as long as the number of atoms we examine is large, because our system is in thermodynamic equilibrium.

§7. *The Expression for the Rate.* Eq. 14.5 tells me how many atoms in the system are in the region A*. To develop a rate theory, I must connect this quantity to the rate of reaction for leaving A. To do this, I need to examine the potential energy in A*. Fig. 14.4 shows the potential energy surface as a function of x for fixed y and z. The position of the dividing surface is indicated by the dashed line and the

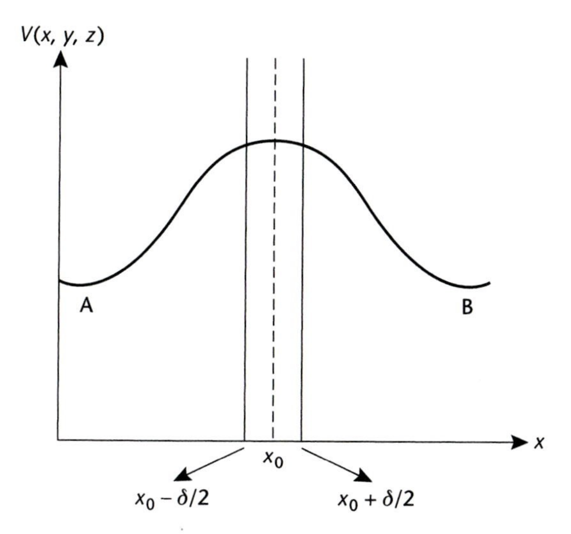

Figure 14.4 The potential energy surface as a function of x (for fixed values of y and z). The ridge is at x_0. The planes $x = x_0 - \delta/2$ and $x = x_0 + \delta/2$ are the borders of the strip that defines the transition complex.

borders of the region A* by solid lines. For different values of y and z, the curves are different, but they all have the same feature: a maximum at x_0 and minima at A and B. This feature is the only thing that matters here.

Next, I take δ to be so small that the potential in the region A* is constant. This means that the force $F_x = -\partial V/\partial x$ is zero within A*. Because of this, a particle that enters A* from the left (the A side) with a velocity component v_x will leave that region with the same velocity v_x. (Mechanics says that the velocity component v_x does not change in a region where the force component F_x, along the x coordinate, is zero.)

This observation makes life easy. A particle that enters the region A* through the surface at $x_0 - \delta/2$ must have $v_x > 0$ (moves in the direction of increasing x) and maintains the same velocity until it exits through the plane at $x_0 + \delta/2$ and penetrates into region B. Once a particle has entered A* from A, it will not get stuck in A* or turn back before reaching B. This means that the number of particles leaving A per unit time, $(dN_A/dt)_-$, is equal to the number of particles leaving A* per unit time, dN^*/dt:

$$\left(\frac{dN_A}{dt}\right)_- = \frac{dN^*}{dt} \tag{14.6}$$

I ignore the particles that leave B to enter A* (which have negative velocity) and exit into A. These particles contribute to the backward rate $(dN_A/dt)_+$. Here I am concerned only with particles that leave A* and have a velocity $v_x > 0$.

Eq. 14.6 is the second important equation in this theory. Unlike Eq. 14.5, this equation is *approximate*. The hidden approximation is this: I assumed that a particle that leaves A* to go into B does not come back for a very long time. This need not be true. A trajectory like the one shown in Fig. 14.5 is possible, but the theory developed here ignores such events. This assumption, of the absence of *recrossing*, will be discussed again later. Here I only mention that experience accumulated through computer simulations shows that in most systems, most trajectories that cross the dividing surface do not recross it. Our approximation is fairly good.

§8. *The Rate to Escape from* A. If I combine Eq. 14.6 with Eq. 14.2, I obtain

$$\left(\frac{dN_A}{dt}\right)_- = \frac{dN^*}{dt} = -k_{A\rightarrow B}N_A \tag{14.7}$$

If I could calculate dN^*/dt, this equation would provide an expression for the rate constant $k_{A\rightarrow B}$. So, let us calculate dN^*/dt.

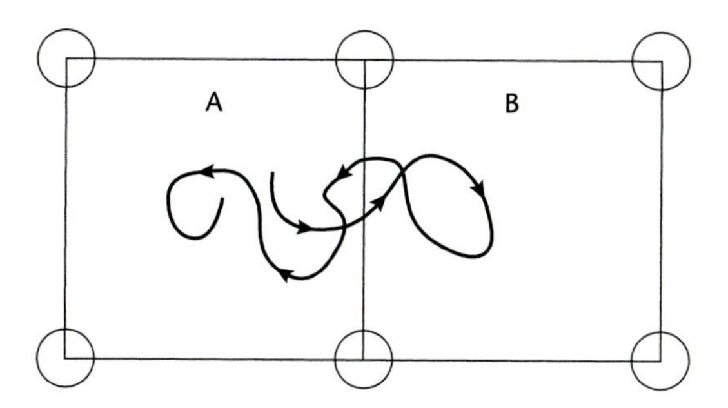

Figure 14.5 A trajectory starting in the cell A, crossing the dividing surface and recrossing soon after arriving in B.

I have already noted that particles that penetrate into A* from A fly across the width δ without change of velocity. How many escape from A*, into B, in unit time? That number is dN^*/dt.

At any given time there are N^* atoms in the region A*. I denote by $p(v_x)\,dv_x$ the probability that such an atom has a velocity between v_x and $v_x + dv_x$. With this notation, the number of atoms having such velocity is $N^*p(v_x)\,dv_x$.

How many of them escape from A* into B? An atom having a velocity v_x travels, in a time dt, the distance $v_x dt$ towards the border with region B. (I am only concerned here with atoms having $v_x > 0$, hence moving towards B.) All the atoms located within a distance $v_x dt$ from the border with B escape from A* in the time interval dt (see Fig. 14.6).

Since within A* the force acting on the atom is zero, there is no preference for a position inside A*. All positions are equally probable. Therefore the probability that an atom in A* is located in the region $[x_0 + \frac{\delta}{2} - v_x dt, x_0 + \frac{\delta}{2}]$ is the width of this region (which is $v_x dt$) divided by the width δ of A*.

Now put it all together: the number of atoms that have a velocity $v_x > 0$ and escape from A* into B in a time dt is

$$N^*p(v_x)\,dv_x \frac{v_x dt}{\delta} \tag{14.8}$$

In kinetics, we do not care what velocity the atoms have when they arrive at B. We only want to know how many atoms arrive at B per unit time. To obtain this value, I integrate the expression in Eq. 14.8 over all positive velocities (atoms with

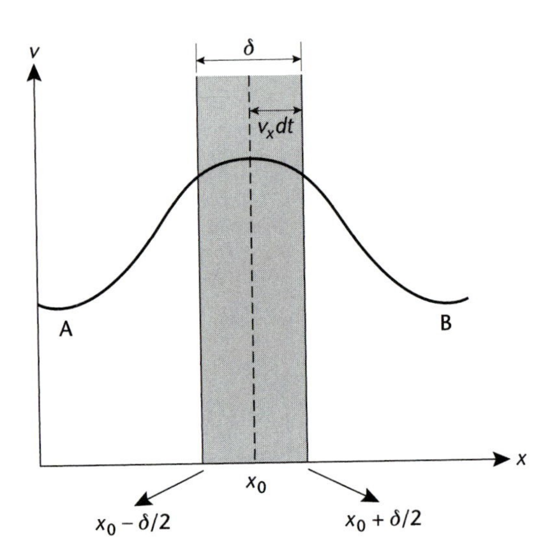

Figure 14.6 All atoms in the gray strip moving from left to right with velocity component v_x will reach region B in the time interval *dt*.

negative velocity move away from B). Therefore, the number of atoms leaving A* to go into B in the time interval dt is

$$dN^* = -\left[\int_0^\infty v_x p(v_x) dv_x\right] \frac{N^* dt}{\delta}$$

The negative sign is necessary because the process we are discussing removes atoms from A* and therefore $dN^* < 0$. The number of atoms leaving A* to go into B per unit time is

$$\frac{dN^*}{dt} = -\frac{N^*}{\delta} \int_0^\infty v_x p(v_x) dv_x \tag{14.9}$$

If I combine Eq. 14.7 with Eq. 14.9, I have

$$\left(\frac{dN_A}{dt}\right)_- = -\frac{N^*}{\delta} \int_0^\infty v_x p(v_x) dv_x$$

Using Eq. 14.5 for N^* gives

$$\left(\frac{dN_{\mathrm{A}}}{dt}\right)_{-} = -\frac{q^* N_{\mathrm{A}}}{\delta q_{\mathrm{A}}} \int_0^{\infty} v_x p(v_x) dv_x \tag{14.10}$$

This is the equation we were looking for!

§9. *The Rate Constant.* Let us stop here and take a deep breath. When we studied chemical kinetics, we used the equation

$$\left(\frac{dN_{\mathrm{A}}}{dt}\right)_{-} = -k_{\mathrm{A}\to\mathrm{B}} N_{\mathrm{A}} \tag{14.11}$$

This was discovered empirically. It was the starting point of chemical kinetics and it has been confirmed by many experiments. It is now widely accepted that the rate of an elementary unimolecular reaction is given by this law. There was, however, no proof that such an equation follows from some fundamental laws. Eq. 14.10 provides the missing proof. The rate dN_{A}/dt given by that equation is proportional to N_{A}. By comparing Eqs 14.10 and 14.11, I obtain

$$k_{\mathrm{A}\to\mathrm{B}} = \frac{q^*}{\delta q_{\mathrm{A}}} \int_0^{\infty} v_x p(v_x)\, dv_x \tag{14.12}$$

With the notation

$$\langle v_x; v_x > 0\rangle \equiv \int_0^{\infty} v_x p(v_x)\, dv_x \tag{14.13}$$

I can rewrite this as

$$k_{\mathrm{A}\to\mathrm{B}} = \frac{q^*}{\delta q_{\mathrm{A}}} \langle v_x; v_x > 0\rangle \tag{14.14}$$

The symbol $\langle v_x; v_x > 0\rangle$ is the mean (average) positive velocity along the x direction. Such a restricted mean is not that unusual. A similar quantity would be the mean grade of all the students whose grade is above the class median.

Exercise 14.1

Make the same type of argument to show that

$$k_{\mathrm{B}\to\mathrm{A}} = \frac{q^*}{\delta q_B}\langle v_x; v_x < 0\rangle \tag{14.15}$$

This problem is a good test of your understanding of the material presented so far.

§10. *Can I Calculate $k_{\mathrm{A}\to\mathrm{B}}$ from Eq. 14.14?* Let us examine Eq. 14.14 from a practical point of view. Can I calculate $k_{\mathrm{A}\to\mathrm{B}}$ from it? I have already pointed out that q^* and q_{A} can be calculated. In some cases, these calculations are complicated, but they can be done.

The average $\langle v_x; v_x < 0\rangle$ is easy to calculate. The atoms crossing the dividing surface are at equilibrium; therefore $p(v_x)dv_x$ is given by Maxwell's distribution:

$$p(v_x)dv_x = \frac{\exp\left[-mv_x^2/2k_BT\right]dv_x}{\int_{-\infty}^{+\infty}\exp\left[-mv_x^2/2k_BT\right]dv_x} \tag{14.16}$$

Here m is the mass of the atom. Therefore

$$\langle v_x; v_x > 0\rangle = \int_0^{\infty} v_x p(v_x)dv_x = \frac{\int_0^{\infty} v_x\exp\left[-mv_x^2/2k_BT\right]dv_x}{\int_{-\infty}^{+\infty}\exp\left[-mv_x^2/2k_BT\right]dv_x} = \sqrt{\frac{k_BT}{2\pi m}} \tag{14.17}$$

The first equality is the definition Eq. 14.13; the second is obtained by using Eq. 14.16; the third follows by performing the integrals. Similar integrals appear often in physics and they are tabulated.[a] They are easily performed by **Mathematica** or **Mathcad**.

Using Eq. 14.17 in Eq. 14.14 gives

$$k_{\mathrm{A}\to\mathrm{B}} = \sqrt{\frac{k_BT}{2\pi m}}\,\frac{q^*}{\delta q_{\mathrm{A}}} \tag{14.18}$$

[a] H.B. Dwight, *Tables of Integrals and Other Mathematical Data,* MacMillan Publishing, New York, 1961

The only unknown quantity in this equation is δ. All I know about δ is that it must be very small. That doesn't pin it down! This expression for the rate is completely useless unless δ disappears from it. You will see later that q^* is proportional to δ, and this takes care of our anxiety.

Exercise 14.2

Convince yourself that the expression on the right in Eq. 14.18 has the correct units.

The Rate Constant $k_{B\to A}$; the Detailed Balance

§11. When we studied chemical kinetics, we derived an equation, called *detailed balance*, that all rate constants must satisfy. I will use it to derive a formula for the rate constant $k_{B\to A}$.

The argument deriving the detailed balance is very simple and I repeat it here. Combining Eqs 14.1, 14.2, and 14.3, we find that the rate of disappearance from A is

$$\frac{dA}{dt} = -k_{A\to B}A + k_{B\to A}B$$

At equilibrium, the concentration no longer changes and $dA/dt = 0$. This means that the equilibrium concentrations A_{eq} and B_{eq} must satisfy

$$0 = -k_{A\to B}A_{eq} + k_{B\to A}B_{eq}$$

This can be rewritten as

$$\frac{k_{A\to B}}{k_{B\to A}} = \frac{B_{eq}}{A_{eq}} \tag{14.19}$$

The right-hand side of Eq. 14.19 is the equilibrium constant K. From Chapter 11 we know that

$$K = \frac{B_{eq}}{A_{eq}} = \frac{q_B}{q_A} \tag{14.20}$$

where q_B and q_A are the partition functions of A and B. Combining Eqs 14.19 and 14.20 gives

$$\frac{k_{\mathrm{A}\to\mathrm{B}}}{k_{\mathrm{B}\to\mathrm{A}}} = \frac{q_B}{q_A}$$

From this and Eq. 14.14 we get

$$k_{\mathrm{B}\to\mathrm{A}} = \frac{k_{\mathrm{A}\to\mathrm{B}}\, q_A}{q_B} = \frac{q^*}{\delta q_B}\langle v_x; v_x > 0\rangle \tag{14.21}$$

Thus, detailed balance allows us to obtain an equation for $k_{\mathrm{B}\to\mathrm{A}}$ from the equation for $k_{\mathrm{A}\to\mathrm{B}}$. Note that this is the formula given in Eq. 14.15, because $\langle v_x; v_x > 0\rangle = \langle v_x; v_x < 0\rangle$.

Exercise 14.3

Show that $\langle v_x; v_x > 0\rangle = \langle v_x; v_x < 0\rangle$.

The Assumptions made by Transition State Theory

§12. *Introduction.* The theory presented here, called *transition state theory,* was introduced in the 1930s by Evans, Polanyi, Wigner, and Eyring. Among these, Eyring was a most vigorous proponent and it is hard to understand why he did not receive a Nobel Prize for it. After all, the rate of chemical reactions and rate theory are the most important subjects in chemistry.

It may be that the Prize eluded Eyring because the theory was impossible to implement conclusively. No one knew how to calculate potential-energy surfaces with any reliability, the computers necessary to study the motion of atoms did not exist, and the theory seemed perfectly useless. Besides, it was plagued by misunderstandings. One of the most damaging was the belief that calculating the number of atoms N^* forming the activated complex by using equilibrium statistical mechanics (as we did in §6, where we used Eq. 14.4) was a crude and out-of-control approximation. It was felt that a molecule engaged in a reaction goes through wild, out of equilibrium states. Some people considered that the theory was just speculation and could not be tested.

Here I examine two aspects of the theory: the erroneous belief that the equilibrium calculation of N^* is approximate and the assumption that recrossing is not likely to be important.

§13. *Should We Use Equilibrium Statistical Mechanics to Calculate N^*?* To answer this question, I need to first remind you of a few things about the system we study. We have N atoms on the surface. Each atom could be in one of the two sites, denoted by A and B.

Each atom is in equilibrium with the solid. This means that its energy can be calculated from *equilibrium* statistical mechanics. Equally important is that the populations N_A and N_B *are also in equilibrium.* This means that the equilibrium constant N_B/N_A can be calculated from equilibrium theory. At the start of our experiment, every quantity we can think of has its equilibrium value.

Now we let the system evolve and do nothing to disturb the equilibrium. We just follow the motion of the particles that were initially in A. I ignore those that were in B as well as those that make it later to an A site (from B).

I denote by $\bar{N}_A(t)$ the number of particles that started in A at time 0 and are still in A at time t. I can find $N_A(t)$ by solving Newton's equations for all particles, on a gigantic computer. It does not matter whether such a computer exists: this is a thought experiment.

What would I observe if I monitor only the atoms that start in A? If I follow a large number of atoms, I will find that $\bar{N}_A(t) = \bar{N}_A(0)\exp[-k_{A\to B}t]$ (this is the solution of $d\bar{N}_A(t)/dt = k_{A\to B}\bar{N}_A(t)$), and I can determine the rate constant $k_{A\to B}$.

Throughout all of this, the system was in equilibrium. The populations $N_A(t)$ and $N_B(t)$ maintained their equilibrium values. You might object that this is not how we do kinetic experiments. In those we start with the isomer A and there is no B. The populations N_A and N_B are not in equilibrium. The essence of chemical kinetics is that the concentrations of the reacting species are not in equilibrium. Kinetics studies how these quantities evolve to equilibrium.

Fair enough, this is a perceptive observation. Now let us compare the two situations: in one sample, all atoms are in A; in the other, the atoms are in A and B, and N_A and N_B have the equilibrium values. I remind you that the atoms in either system do not interact with each other. There is no way in which they can signal to each other: hey guys, according to equilibrium theory there are too many of us in A, half of us should jump into B. Each atom in A jumps into B as if it is alone in the universe. An atom in A in the sample where N_A and N_B are in equilibrium jumps to B with the same rate as the one in the sample where they all start in A.

This analysis tells me that it is perfectly legitimate to calculate the jumping rate constant $k_{A \to B}$ in a system in which N_A and N_B are in equilibrium, and use it for describing a system in which N_A and N_B are not in equilibrium.

§14. *Each Atom is in an Equilibrium State at All Times.* Now that we feel secure in performing kinetic studies in a system that is at equilibrium, I can make the main point. We start the system at equilibrium and we do nothing to disturb it. Therefore, each atom is at equilibrium at all times, including that moment when it crosses the energy ridge separating A from B. This means that the probability that the atom reaches a configuration C very close to the ridge is proportional to $\exp[-V(C)/k_B T]$, as given by equilibrium statistical mechanics. This is the formula used in §6, which led to the contested conclusion, Eq. 14.4. We have learned that Eq. 14.4 is correct; it does not involve any approximation.

§15. *If Everything is in Equilibrium, Why Do We Have a Reaction?* The one-word answer is fluctuations. Since you have not studied probability and statistics, I will spend some time explaining fluctuations to you.

Let us start with coin flipping. Everybody knows that the probability of getting heads with a fair coin is $\frac{1}{2}$, and so is the probability of getting tails. Does this mean that if I flip a coin ten times I will get heads, tails, heads, tails, ...? Or that I will get five heads and five tails? No. If I flip the coin N times and count N_h heads in the result, probability theory says that

$$\frac{N_h}{N} \to \frac{1}{2} \tag{14.22}$$

if N is very large. This rule does not hold for ten flips.

Now let us get back to our ten flips. Denote by $\langle N_h \rangle$ and $\langle N_t \rangle = N - \langle N_h \rangle$ the numbers of heads and of tails that I would get if I use the formula in Eq. 14.22. For $N = 10$, $\langle N_h \rangle = \frac{1}{2} \times 10 = 5$ and $\langle N_t \rangle = 5$. Denote by N_h and N_t the numbers of heads and tails actually observed in an experiment of ten flips. If you perform several such experiments, you will discover that the values of N_h given by each experiment (of ten flips) differ from one experiment to another. I am not saying that these values are never the same; I am saying that they don't have to be the same. Moreover, N_h can be different from $\langle N_h \rangle$ — it doesn't have to be, but it often is.

The difference

$$N_h - \langle N_h \rangle$$

is called a fluctuation of the quantity N_h. It measures by how much the value of N_h obtained in an experiment differs from the value predicted by probability theory. For example, if $N_h = 6$ and you performed $N = 10$ flips, then you have observed a fluctuation: $N_h - \langle N_h \rangle = 6 - 5 = 1$.

If the number of flips N is very large, then $|N_h - \langle N_h \rangle| \ll N_h$ and the fluctuation is negligible. It is always there, but if you use an instrument to measure N_h and its accuracy is to three decimal places, you'll be unable to tell that $N_h \neq \langle N_h \rangle$ when N is very large.

§16. *Density Fluctuations.* Now let us look at an example from physics. You know from thermodynamics (and common sense) that the density of a liquid is the same at every point. How do we know that? We take samples of 1 cm^3 of liquid from two places in a beaker and weigh the samples. They have "exactly" the same weight. "Exactly" here means within the accuracy of the measurement.

Now you go a step further. The two samples taken have the same volume (1 cm^3) and the same weight, so they must have the same number of molecules. Good logic, bad physics. What if one sample has one molecule more than the other? Could you tell? No. There is no method of measurement that can tell you that. How about 100 extra molecules? Still no!

The statement that the density in two places in a liquid is the same, and that equal volumes in the liquid contain the same number of molecules, is based on the crudity of measurement. We are large and molecules are small, and, until very recently, we were only able to make measurements on large numbers of molecules. It is as if we are allowed to make coin-flipping experiments only with an enormous number of flips and, at the same time, we are unable to keep track accurately of the number of heads.

Now imagine that we are able to observe a cube with a size of 30 Å. Since molecular motion is very disorganized, there is no reason to have the same number of molecules in cubes located at different places in the liquid. Nor must the number of molecules in the same cube be the same at different times. The number of molecules in a small cube fluctuates around the value predicted by probability theory, which says that two cubes of equal volume contain an equal number of molecules.

This seems plausible, but how do we prove it? Optics comes to the rescue. By solving Maxwell's equation for light propagation, one can show that if the density of a liquid is the same at every point, then light will go through the liquid in a straight line. This means that the detector D in Fig. 14.7 will detect no light. This

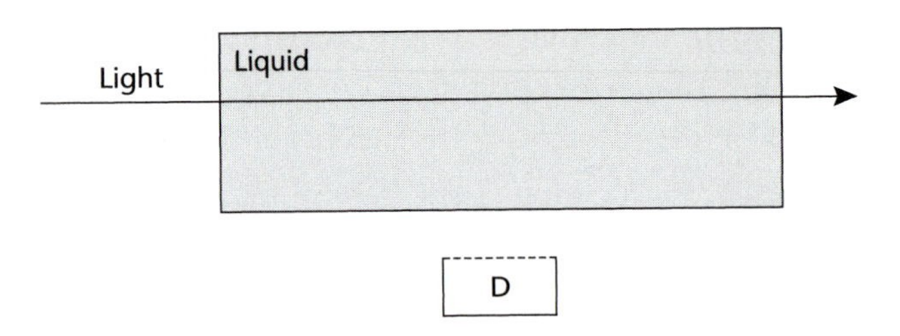

Figure 14.7 In the absence of fluctuations light travels through the sample in a straight line and does not reach the detector D.

is not what happens. What went wrong? Light propagation is controlled by the refractive index, which depends on the density of the liquid. We have assumed that the density is the same at all points in the liquid. This makes every point in the liquid have the same refractive index. This is why light should travel through the liquid in a straight line.

What happens if fluctuations exist? The density and the refractive index will then differ from place to place. These deviations from the mean density (fluctuations) take place only in small volumes. But light is very sensitive to them and they cause some photons to be scattered away from the forward direction of propagation. The detector in Fig. 14.7 will detect this scattered light.

If we take fluctuations into account in solving Maxwell's equations for light propagation, we can calculate precisely how the intensity of the scattered light depends on the position of the detector and on the light frequency. These predictions of theory have been tested and agree well with experiment. Fluctuations are real and have detectable effects.

§17. *Energy Fluctuations and Chemical Reactions.* The energy measured in thermodynamics is the average energy of the molecules in the system. By dividing it by the number of molecules, I obtain the mean energy $\langle E \rangle$ per molecule.

This quantity is very similar to the number of heads $\langle N_h \rangle$ in the coin-flipping experiment. If I were to measure the energy ε of one molecule in a liquid or a gas, it is very likely that I will find a value that differs from $\langle E \rangle$. Suppose we measure ε and then leave the molecule alone (in the liquid) for a while and later measure ε again it is very likely that the two measurements will give different values. If I do a huge number N of such measurements, and I get the values $\varepsilon_1, \varepsilon_2, \ldots, \varepsilon_N$, then

$$\frac{\varepsilon_1 + \varepsilon_2 + \cdots + \varepsilon_N}{N} = \langle E \rangle$$

The energy ε of a molecule fluctuates. It is not hard to understand why. The motion of the molecules is very chaotic and a given molecule suffers many collisions with its neighbors. These collisions change the internal (vibrational) energy of the molecule. The energy fluctuates and occasionally the molecule acquires enough energy in the reaction coordinate to undergo a chemical reaction.

These fluctuations are processes that take place in a system that is in equilibrium. Their probability can be calculated using equilibrium statistical mechanics. This is why the probability that an energy fluctuation forms an activated complex is given by Eq. 14.4.

Briefly stated, chemical reactions take place because an equilibrium fluctuation increases the energy of the reaction coordinate above the activation energy. Change requires above-average energy and it is slow because such a surge in energy is rare.

§18. *Recrossing.* We know now that Eq. 14.4 is not an approximation. Does that mean that transition state theory is exact? No. We assumed that a trajectory that crosses the dividing surface, from A to B, does not go back soon to A, but settles in B and spends a fairly long time there; the atom is a resident of B, not a tourist visiting it. Often this is a good approximation, but strictly speaking it is not true in all cases and sometimes it can produce large errors. In what follows, I examine the physics behind this no recrossing assumption.

Why do we make it? You may have noticed that transition state theory does not require me to calculate reactive trajectories. This is because I have assumed that no recrossing takes place. Indeed, if recrossing is allowed, I should count trajectories like those shown in Fig. 14.8a and c, but I should not count ones like that in Fig. 14.8b.

To decide if a trajectory should be counted, I must run it, follow it, and see whether the chemisorbed atom settles in A or in B. Transition state theory assumes that they all settle in B, so I don't have to follow them to see if they do. This saves a lot of computing. However, it also makes me count the trajectory in Fig. 14.8b as reactive, although it is not. That trajectory contributes to the average velocity $\langle v_x; v_x > 0\rangle$ and it shouldn't. The value of $\langle v_x; v_x > 0\rangle$ calculated by transition state theory is larger than the correct value, and it is equal to the correct value only if no recrossing takes place. Because the rate constant is proportional to $\langle v_x; v_x > 0\rangle$ (see Eq. 14.14), we have

$$k_{\text{TST}} \geq k_{\text{exact}}.$$

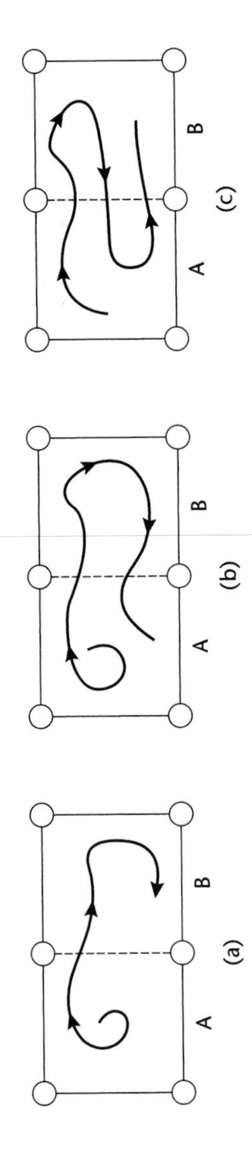

Figure 14.8 (a) and (c) show reactive trajectories whose contribution to the rate constant must be taken into account. (b) shows a non-reactive trajectory, whose contribution should not be counted.

The rate constant calculated by transition state theory is larger than or equal to the exact rate constant.

§19. *Why Does the No-Recrossing Approximation Work?* Let us look more closely at our example to see why most trajectories do not recross. Remember that, to be able to cross the ridge, a reactive trajectory must have high energy in the reaction coordinate. Once it enters into the region B, it continues moving away from the dividing surface. Then it interacts with the atoms surrounding cell B, it is attracted to or repelled by them, and it changes direction. Sooner or later, it will turn around and move towards A. The key question is whether it has enough energy in the reaction coordinate to recross the dividing surface. If it does, it violates the no-recrossing requirement. If it does not, all is well with transition state theory.

There is a time τ needed for the atom to change its direction and make v_x negative (to move towards A). Two things happen in this time. Some of the energy in the x-coordinate (the reaction coordinate) is transferred to the y and z directions. For example, the atom reaching B might start oscillating perpendicular to the surface. The energy of this motion is taken from the motion in the x-direction. Such an atom cannot recross, since it no longer has enough energy in the reaction coordinate.

Furthermore, the atom collides with the surface atoms and exchanges energy with them. Since our atom has high energy (it just crossed the barrier) and the atoms in the solid are likely to have a near-average energy (most atoms do), energy is transferred from the energetic atom to the sluggish ones. The solid gains energy and the chemisorbed atom loses it.

Thus, changing its energy from the reaction coordinate into other coordinates and robbed of energy by the solid's atoms, our exhausted traveler settles down in B. It does not recross back into A.

This situation is reflected by equilibrium statistical mechanics. In Chapter 12, you learned that an atom can maintain an energy (kinetic or potential) higher than average only for a very short time. If this time is shorter than the time it takes the atom to turn around to head towards A, the atom does not have enough energy to recross.

Our atom is like a tourist who leaves A with a one-way ticket and a finite amount of money. He arrives at his destination, spends too much money for meals and entertainment, and does not have enough money for a return ticket. He settles there, takes a job, saves money, and goes back a year later. That later trip is no longer a recrossing. He has lived too long in B (a time of order $1/k_{B\to A}$) and now his return counts as a reaction B→A.

Supplement 14.1 The Dividing Surface in General

Because of the high symmetry of the system considered in this chapter, the ridge dividing A from B is a straight line and the dividing surface is a plane. This is not true for most reactions. In general the ridge is an abstract line in a multidimensional space, because many coordinates are needed to specify its position. In all cases, the dividing surface is defined to intersect the ridge, on the potential-energy surface, separating the reactants and the products. It must be such that any reactive trajectory must pass through it on the way from reactant minimum to product minimum. The ridge need not be a straight line and the dividing surface need not be a plane.

Let us see how this works out for the reaction HCN → CNH. When the molecule evolves from reactant to product, the hydrogen atom must swing around the CN group (Fig. 14.9). To simplify matters, let us assume that the C–N bond length does not change when the hydrogen atom swings around, in its trip from HCN to CNH. This is not true, but let's pretend it is, for a while. Then, the reactive trajectories and the shape of the potential-energy surface are described by two variables: r and ϕ. In other words, $V = V(r, \phi)$. $V(r, \phi)$ has two minima: one for $\phi = 0$ and $r = r_1$ (the length of the C–H bond in HCN); the other for $\phi = 180°$ and $r = r_2$ (the length of the C–H bond in CNH). Of the two coordinates, ϕ is the more convenient for describing the reaction, which is a rotation of H around CN. So as ϕ changes from 0° (HCN) to 180° (CNH), $V(r, \phi)$ must go through a maximum value. At that maximum,

$$\frac{\partial V(r,\phi)}{\partial \phi} = 0 \qquad (14.23)$$

This condition provides a relationship between ϕ and r, which I denote by $\phi = f(r)$. I obtain it by solving Eq. 14.23 to find how ϕ depends on r.

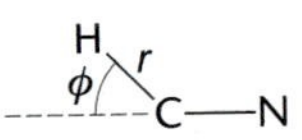

Figure 14.9 The reaction HCN → CNH is described by two variables, the angle ϕ and the C–H bond length r.

$\phi = f(r)$ is the equation of the ridge, in the polar coordinates ϕ and r. It is very unlikely that this is a straight line.

Exercise 14.4

Use Lennard-Jones interactions between H and C and between H and N, calculate the potential-energy of the HCN system, and determine numerically where the ridge is positioned. Keep the C–N distance constant in one calculation. Then, in another calculation, let the C–N distance vary. Compare the results. This is not a good potential-energy for the HCN molecule, but you'll learn quite a bit from this exercise.

Of course, the approach is crude. There is no reason why the carbon–nitrogen bond will stay fixed when the hydrogen atom moves around. The potential-energy depends on ϕ, r, and R, where R is the C–N distance. So when I look for the ridge, I must let R vary also. As the science becomes more realistic, the life of the scientist becomes more complicated. However, by using abstract mathematics and computers, we can find the ridge and define a dividing surface, for very complex systems.

Why did I get into this? To warn you that the picture you see in the books, where the reaction is described by one coordinate, the ridge is one point, and the dividing surface is a line, is very misleading. There is no such reaction. In all cases of interest in practice, the reaction coordinate is a complicated combination of the coordinates of several atoms. Even the simple example of the migration of a chemisorbed atom is simple only because we have held the solid's atoms fixed. In reality, when an atom moves from A to B and crosses the ridge, the solid's atoms close to the chemisorbed one will also move. Their motion must be included when we calculate the position of the ridge and it is part of the reaction coordinate.

15

TRANSITION STATE THEORY: CALCULATING THE RATE CONSTANT

Introduction

§1. *A Survey of the Procedure for Calculating* $k_{\mathrm{A}\to\mathrm{B}}$. The equation

$$k_{\mathrm{A}\to\mathrm{B}} = \sqrt{\frac{k_B T}{2\pi m}} \frac{q^*}{\delta q_{\mathrm{A}}} \tag{15.1}$$

derived in the previous chapter (see Eq. 14.18) gives us the means for calculating the rate constant $k_{\mathrm{A}\to\mathrm{B}}$. To implement it, we need to calculate the partition functions q_{A} and q^*.

Most of the time the motion of the reactant A is a small-amplitude vibration around the point $\mathbf{r}_A = \{x_{\mathrm{A}}, y_{\mathrm{A}}, z_A\}$ where the potential energy has a minimum. Because of this, we can calculate the partition function q_{A} by approximating the potential energy $V(\mathbf{r})$ with its second-order Taylor expansion (power series expansion) in terms of small displacements around the point $\mathbf{r}_{\mathrm{A}}$, at which $V(\mathbf{r})$ has a minimum.

This is the harmonic approximation, that you encountered in Chapter 9, when you studied vibrational partition function.

After we make the harmonic approximation, we use the second-order derivatives of $V(\mathbf{r})$ with respect to x, y, and z (the *force constants*) to calculate the vibrational frequencies. Once we know these we can calculate the vibrational partition functions to obtain a good approximation of q_A.

The partition function for the transition complex is more subtle. While located in the region A* (see Fig. 14.4), the particle executes a *translational motion* in the x direction (the *reaction coordinate*) and oscillates in the y and z directions. The partition function q^* is the product of a translational and of two vibrational partition functions. The latter are calculated by using the harmonic approximation for y and z variables around the transition state position $\mathbf{r}_T$.

The translational partition function is proportional to δ and because of this δ disappears from Eq. 15.1.

These procedures give us a formula for $k_{A\to B}$, which is used for numerical calculations in Workbook SM15.1, where we examine the Arrhenius formula and the isotope effect. The implementation of these ideas is described in what follows.

The Partition Function q_A

§2. *The Harmonic Approximation.* I could use a numerical method to calculate q_A. Since you have not learned how to do that, I will make an approximation that is fairly accurate in the case of the migration of a chemisorbed atom. Because the potential energy is a bowl with the minimum at the point $\mathbf{r}_A = \{x_A, y_A, z_A\}$, the motion of the atom in cell A is an oscillation around the center of the cell.

Most of the time the particle is in the vicinity of $\mathbf{r}_A$. Because of this I can study this motion by using an approximate representation of $V(x, y, z)$ that is accurate only when the atom is close to $\mathbf{r}_A$.

As you have learned in calculus, a few terms in a Taylor series expansion around $\mathbf{r}_A$ provides such an approximation. This expansion is shown below.

$$V(x,y,z) \cong V(x_A,y_A,z_A) + \left(\frac{\partial V}{\partial x}\right)_A (x - x_A) + \left(\frac{\partial V}{\partial y}\right)_A (y - y_A) + \left(\frac{\partial V}{\partial z}\right)_A (z - z_A) + \frac{1}{2}\left(\frac{\partial^2 V}{\partial x^2}\right)_A (x - x_A)^2 + \frac{1}{2}\left(\frac{\partial^2 V}{\partial y^2}\right)_A (y - y_A)^2$$

$$+\frac{1}{2}\left(\frac{\partial^2 V}{\partial z^2}\right)_A (z-z_A)^2 + \left(\frac{\partial^2 V}{\partial x \partial y}\right)_A (x-x_A)(y-y_A)$$
$$+\left(\frac{\partial^2 V}{\partial x \partial z}\right)_A (x-x_A)(z-z_A) + \left(\frac{\partial^2 V}{\partial y \partial z}\right)_A (y-y_A)(z-z_A) \quad (15.2)$$

The subscript A in $(\partial V/\partial x)_A$, etc., indicates that the partial derivative is evaluated at the point $\mathbf{r}_A = \{x_A, y_A, z_A\}$. Because $\mathbf{r}_A$ corresponds to a minimum of $V(x, y, z)$, the first-order derivatives $(\partial V/\partial x)_A$, $(\partial V/\partial y)_A$, and $(\partial V/\partial z)_A$ are all equal to zero. Therefore Eq. 15.2 becomes

$$V(x,y,z) \simeq V(x_A,y_A,z_A) + \frac{1}{2}\left(\frac{\partial^2 V}{\partial x^2}\right)_A (x-x_A)^2$$
$$+\frac{1}{2}\left(\frac{\partial^2 V}{\partial y^2}\right)_A (y-y_A)^2 + \frac{1}{2}\left(\frac{\partial^2 V}{\partial z^2}\right)_A (z-z_A)^2$$
$$+\left(\frac{\partial^2 V}{\partial x \partial y}\right)_A (x-x_A)(y-y_A) + \left(\frac{\partial^2 V}{\partial x \partial z}\right)_A (x-x_A)(z-z_A)$$
$$+\left(\frac{\partial^2 V}{\partial y \partial z}\right)_A (y-y_A)(z-z_A) \quad (15.3)$$

§3. *The Physical Meaning of the Terms in the Harmonic Expansion.* The energy

$$V_A \equiv V(x_A, y_A, z_A) = -1.6794 \text{ eV} \quad (15.4)$$

is the lowest energy the atom can have in cell A. It is the local minimum of $V(x, y, z)$, the bottom of the bowl. The numerical value was calculated in Workbook SM15.1. In physical chemistry (and physics), the second derivatives of the potential energy are called the *force constants* and are often denoted by k_{xx}, k_{xy}, etc.

Exercise 15.1

Calculate the forces F_x, F_y, F_z acting on the particle, when the potential energy surface is given by Eq. 15.3. Show that these forces pull the particle towards $\mathbf{r}_A$ and that they are proportional to the force constants. *Hint.* Use $F_x = -(\partial V/\partial x)$, $F_y = -(\partial V/\partial y)$, $F_z = -(\partial V/\partial z)$.

The numerical values of the force constants were calculated in Cell 2 of Workbook SM15.1. They are

$$k_{xx} \equiv \frac{1}{2}\left(\frac{\partial^2 V}{\partial x^2}\right)_A = 1.517 \text{ eV/Å}^2 \tag{15.5}$$

$$k_{yy} \equiv \frac{1}{2}\left(\frac{\partial^2 V}{\partial y^2}\right)_A = 1.514 \text{ eV/Å}^2 \tag{15.6}$$

$$k_{zz} \equiv \frac{1}{2}\left(\frac{\partial^2 V}{\partial z^2}\right)_A = 3.424 \text{ eV/Å}^2 \tag{15.7}$$

In the same Workbook , I found that

$$k_{xy} \equiv \left(\frac{\partial^2 V}{\partial x \partial y}\right)_A = 1.766 \times 10^{-15} \text{ eV/Å}^2$$

$$k_{xz} \equiv \left(\frac{\partial^2 V}{\partial x \partial z}\right)_A = -0.04457 \text{ eV/Å}^2$$

$$k_{yz} \equiv \left(\frac{\partial^2 V}{\partial y \partial z}\right)_A = -7.412 \times 10^{-15} \text{ eV/Å}^2$$

If you look at Fig. 13.4, you will see that the x and y directions are not quite equivalent; there are more atoms in the horizontal direction than in the vertical direction. Nevertheless, the fact that k_{xz} and k_{yz} are so different surprises me. I could find no error in my calculation, so I will use these values. Furthermore, the cross-force constants k_{xy}, k_{xz}, and k_{yz} are much smaller than k_{xx}, k_{yy}, and k_{zz} and, I neglect them.

After this, the harmonic approximation to the potential-energy surface around the point A $= \{x_A, y_A, z_A\} = \{4.2032, 4.2, 2.0837\}$ Å (see Cell 4 of Workbook SM15.1 for the numerical values) is

$$\begin{aligned} V(x,y,z) \cong &-1.6754 + 1.5138\,(x - 4.2032)^2 \\ &+ 1.5178\,(y - 4.2)^2 + 3.4239\,(z - 2.0837)^2 \end{aligned} \tag{15.8}$$

In this formula, x, y, and z are in Ångstroms and the energy is in eV.

§4. *The Vibrational Frequencies.* It is a simple problem in mechanics to show that the motion of a particle having the potential given by Eq. 15.8 is that of three independent (i.e. uncoupled) harmonic oscillators, having the frequencies

$$\begin{aligned}\omega_x &= \sqrt{\frac{k_{xx}}{m}} = 1.204 \times 10^{14}\sqrt{\frac{1}{m}}\ \mathrm{s}^{-1}\\ \omega_y &= \sqrt{\frac{k_{yy}}{m}} = 1.206 \times 10^{14}\sqrt{\frac{1}{m}}\ \mathrm{s}^{-1}\\ \omega_z &= \sqrt{\frac{k_{zz}}{m}} = 1.810 \times 10^{14}\sqrt{\frac{1}{m}}\ \mathrm{s}^{-1}\end{aligned} \tag{15.9}$$

The last equalities have been obtained by using the numerical values of the force constants (see Cell 3 of Workbook SM15.1). The mass m in these equations is in atomic mass units (e.g. m for oxygen is 16).

Exercise 15.2

Use Newton's equations

$$m\frac{d^2x}{dt^2} = -\frac{dV}{dx};\quad m\frac{d^2y}{dt^2} = -\frac{\partial V}{\partial y};\quad m\frac{d^2z}{dt} = -\frac{\partial V}{\partial t}$$

to show that:

$$\begin{aligned}x(t) - x_A &= C_1 \cos\left(t\sqrt{\frac{k_{xx}}{m}}\right) + C_2 \sin\left(t\sqrt{\frac{k_{xx}}{m}}\right)\\ y(t) - y_A &= C_3 \cos\left(t\sqrt{\frac{k_{yy}}{m}}\right) + C_4 \sin\left(t\sqrt{\frac{k_{yy}}{m}}\right)\\ z(t) - z_A &= C_5 \cos\left(t\sqrt{\frac{k_{zz}}{m}}\right) + C_6 \sin\left(t\sqrt{\frac{k_{zz}}{m}}\right)\end{aligned}$$

where $C_1, \ldots, C_6$ are constants that depend on the initial values of $x(t)$, $y(t)$, $z(t)$ and $dx(t)/dt$, $dy(t)/dt$, $dz(t)/dt$. Because $x(t)$ does not depend on $y(t)$ and $z(t)$, we say that the motion in direction x is decoupled from that in directions y and z. The motion in any one direction is decoupled from that in the other two directions.

Note that the frequency of the motion is $\sqrt{k_{xx}/m}$, etc.

Show that this is not true if the cross-force constants k_{xy}, k_{xz} or k_{zy} are not negligible.

---

§5. *The Partition Function* q_A. From quantum mechanics, I know that the energy of these three independent oscillators is

$$E_{n,m,k} = V_A + \hbar\omega_x(n + \tfrac{1}{2}) + \hbar\omega_y(m + \tfrac{1}{2}) + \hbar\omega_z(k + \tfrac{1}{2})$$

where $n = 0, 1, 2, \ldots$; $m = 0, 1, 2, \ldots$; $k = 0, 1, 2, \ldots$. Here $V_A \equiv V(x_A, y_A, z_A)$ is the minimum potential energy.

The partition function is therefore

$$q_A = \sum_{n=0}^{\infty}\sum_{m=0}^{\infty}\sum_{k=0}^{\infty} \exp\left[-\left(V_A + \hbar\omega_x(n + \tfrac{1}{2}) + \hbar\omega_y(m + \tfrac{1}{2}) + \hbar\omega_z(k + \tfrac{1}{2})\right)/k_B T\right]$$

$$= \exp\left[-\frac{V_A + \frac{1}{2}\hbar\omega_x + \frac{1}{2}\hbar\omega_y + \frac{1}{2}\hbar\omega_z}{k_B T}\right]$$

$$\times \sum_{n=0}^{\infty}\exp\left[-\frac{n\hbar\omega_x}{k_B T}\right]\sum_{m=0}^{\infty}\exp\left[-\frac{m\hbar\omega_y}{k_B T}\right]\sum_{k=0}^{\infty}\exp\left[-\frac{k\hbar\omega_z}{k_B T}\right] \tag{15.10}$$

We calculated the sums in this equation in Chapter 9, when we discussed the vibrational partition function of a diatomic molecule, and obtained (see Eq. 9.35)

$$\sum_{n=0}^{\infty}\exp\left[-\frac{n\hbar\omega}{k_B T}\right] = \frac{1}{1 - \exp\left[-\frac{\hbar\omega}{k_B T}\right]} \tag{15.11}$$

Using Eq. 15.11 in Eq. 15.10 leads to

$$q_A = \frac{\exp\left[-\frac{V_A}{k_B T}\right]\exp\left[-\frac{(\hbar\omega_x + \hbar\omega_y + \hbar\omega_z)}{2k_B T}\right]}{\left(1 - \exp\left[-\frac{\hbar\omega_x}{k_B T}\right]\right)\left(1 - \exp\left[-\frac{\hbar\omega_y}{k_B T}\right]\right)\left(1 - \exp\left[-\frac{\hbar\omega_z}{k_B T}\right]\right)} \tag{15.12}$$

There are three kinds of contributions to q_A. The term

$$\exp\left[-\frac{V_A}{k_B T}\right] \tag{15.13}$$

contains the minimum energy V_A. This term is proportional to the probability that the particle is at $\mathbf{r}_A$. The term

$$\exp\left[-\frac{\hbar\omega_x + \hbar\omega_y + \hbar\omega_z}{2k_B T}\right] \tag{15.14}$$

is a contribution from the *zero point energy* of the oscillators. The term

$$\frac{1}{\left(1-\exp\left[-\frac{\hbar\omega_x}{k_B T}\right]\right)\left(1-\exp\left[-\frac{\hbar\omega_y}{k_B T}\right]\right)\left(1-\exp\left[-\frac{\hbar\omega_z}{k_B T}\right]\right)} \tag{15.15}$$

is a contribution from the three independent vibrational motions. The vibrational frequencies ω_x, ω_y and ω_z are given by Eq. 15.10.

Note that if we know $V(x, y, z)$ then we can calculate q_A. You see here one of the many reasons why physical chemists are so keen on developing methods for accurately calculating potential-energy surfaces. ω_x, ω_y and ω_z depend on the mass of the particle. This dependence is important when we study the isotope effect on the rate constant (see §15 and §16).

The Partition Function q^*

§6. *The Motion of the Atom in the Region* A^*. In the x direction, the particle moves freely in a box of width δ. For that degree of freedom, the partition function is

$$q_x^* = \left(\frac{2\pi m k_B T}{h^2}\right)\delta \tag{15.16}$$

This is the partition function of a particle on a one-dimensional box of width δ (see Chapter 5, §3). Here h is the "old" Planck constant $h = 6.6262 \times 10^{-27}$ erg s.

In §19 of Chapter 13, we studied the shape of the potential energy surface around the transition state point (the saddle point) $\mathbf{r}_T = \{x_T, y_T, z_T\}$. There we showed that the surface has a minimum at $\mathbf{r}_T$, with respect to the variables y and z. This means that the motion of the particle in these two directions is oscillatory and we can deal with it by making the harmonic approximation.

This gives (see §2 for a more detailed discussion of this approximation):

$$V(x_T, y, z) = V(x_T, y_T, z_T) + \frac{1}{2}\left(\frac{\partial^2 V}{\partial y^2}\right)_{TS}(y - y_T)^2 + \frac{1}{2}\left(\frac{\partial^2 V}{\partial z^2}\right)_{TS}(z - z_T)^2 + \left(\frac{\partial^2 V}{\partial y \partial z}\right)_{TS}(y - y_T)(z - z_T)$$

The first derivatives do not appear because they are equal to zero: $V(x, y, z)$ has a minimum with respect to y and to z at the saddle point $\{x_T, y_T, z_T\}$. Numerical calculations performed in Cell 5 of Workbook SM15.1 give

$$\begin{aligned} k_{yy}^T &= \frac{1}{2}\left(\frac{\partial^2 V}{\partial y^2}\right)_T = 1.016 \text{ eV/Å}^2 \\ k_{zz}^T &= \frac{1}{2}\left(\frac{\partial^2 V}{\partial z^2}\right)_T = 3.265 \text{ eV/Å}^2 \\ k_{yz}^T &= \left(\frac{\partial^2 V}{\partial y \partial z}\right)_T = 3.515 \times 10^{-15} \text{ eV/Å}^2 \end{aligned} \tag{15.17}$$

The mixed derivative is practically zero, therefore the oscillation in the y direction is decoupled from that in the z direction.

The potential energy surface around the transition state is (Cell 5 of Workbook SM15.1).

$$V(x_T, y, z) = -1.224 + 1.016(y - 4.2)^2 + 3.265(z - 2.461)^2 \tag{15.18}$$

In this equation, y and z are in Å and the result is in eV. The numerical value of $V(x_T, y_T, z_T)$ is -1.22405 eV and it was calculated in Workbook SM15.1. The coordinates of the transition state are $\mathbf{r}_T = \{x_T, y_T, z_T\} = \{5.6, 4.2, 2.461\}$ (see Cell 5 of Workbook SM15.1).

§7. *The Vibrational Frequencies in the Transition State.* You have learned in mechanics that a particle with a harmonic potential energy surface, such as that given by Eq. 15.18, undergoes harmonic oscillations in the y and z directions with frequencies

$$\omega_y^T = \sqrt{\frac{k_{yy}^T}{m}} \quad \text{and} \quad \omega_z^T = \sqrt{\frac{k_{zz}^T}{m}} \tag{15.19}$$

To calculate these frequencies, we need to pay attention to units. For k_{yy}^T and k_{zz}^T, use erg/cm^2 (convert from the eV/Å^2 values in Eq. 15.18), and use grams for mass. If k_{yy}^T and k_{zz}^T have the values given in Eq. 15.18, then (see Cell 6 of Workbook SM15.1)

$$\omega_y^T = 9.864 \times 10^{13} \sqrt{\frac{1}{m}}\ \mathrm{s}^{-1} \tag{15.20}$$

$$\omega_z^T = 1.768 \times 10^{14} \sqrt{\frac{1}{m}}\ \mathrm{s}^{-1} \tag{15.21}$$

Here m is the mass of the particle in au.

§8. *The Partition Function q^*.* The motion of the particle in the A* region is that of a particle in a box in the x direction and of independent harmonic oscillators in the y and z directions. Therefore, the partition function is a product of the translational partition function in the x direction (q_x^* given by Eq. 15.16) and the vibrational partition functions for the oscillatory motions in y and z directions. This gives

$$q^* = \left\{ \delta \sqrt{\frac{2\pi m k_B T}{h^2}} \right\} \exp\left[-\frac{V_T}{k_B T}\right] \exp\left[-\frac{\hbar\omega_z^T + \hbar\omega_y^T}{2k_B T}\right] \times \frac{1}{1-\exp\left[-\frac{\hbar\omega_y^T}{k_B T}\right]} \frac{1}{1-\exp\left[-\frac{\hbar\omega_z^T}{k_B T}\right]} \tag{15.22}$$

The term in the curly braces is the translational contribution. The first exponential is a contribution that is proportional to the probability that the particle reaches the transition state. The next exponential comes from the zero point energy. The remaining two terms represent contributions from the vibrational motion.

The Equation for the Rate Constant

§9. I can now write out the formula for the rate constant, in all its glory. The general result was (Eq. 15.1)

$$k_{\mathrm{A}\to\mathrm{B}} = \sqrt{\frac{k_B T}{2\pi m}} \frac{q^*}{\delta q_{\mathrm{A}}} \tag{15.23}$$

Using, in this equation, Eq. 15.22 for q^* and Eq. 15.10 for q_A gives

$$k_{A\to B} = \sqrt{\frac{k_B T}{2\pi m}} \times \frac{1}{\delta}\left[\delta\sqrt{\frac{2\pi m k_B T}{h^2}}\right]$$
$$\times \exp\left[-\frac{V_T - V_A}{k_B T}\right] \times \exp\left[-\frac{\hbar(\omega_z^T + \omega_y^T - \omega_x - \omega_y - \omega_z)}{k_B T}\right]$$
$$\times \frac{\left(1-\exp\left[-\frac{\hbar\omega_x}{k_B T}\right]\right)\left(1-\exp\left[-\frac{\hbar\omega_y}{k_B T}\right]\right)\left(1-\exp\left[-\frac{\hbar\omega_z}{k_B T}\right]\right)}{\left(1-\exp\left[-\frac{\hbar\omega_y^T}{k_B T}\right]\right)\left(1-\exp\left[-\frac{\hbar\omega_z^T}{k_B T}\right]\right)}$$
$$= \frac{k_B T}{h} \times \exp\left[-\frac{V_T - V_A + \hbar(\omega_z^T + \omega_y^T - \omega_x - \omega_y - \omega_z)}{k_B T}\right]$$
$$\times \frac{\left(1-\exp\left[-\frac{\hbar\omega_x}{k_B T}\right]\right)\left(1-\exp\left[-\frac{\hbar\omega_y}{k_B T}\right]\right)\left(1-\exp\left[-\frac{\hbar\omega_z}{k_B T}\right]\right)}{\left(1-\exp\left[-\frac{\hbar\omega_y^T}{k_B T}\right]\right)\left(1-\exp\left[-\frac{\hbar\omega_z^T}{k_B T}\right]\right)} \qquad (15.24)$$

§10. *A Review of the Equation for $k_{A\to B}$ (Eq. 15.24).* The term $k_B T/h$ has units of s^{-1} and gives $k_{A\to B}$ the correct units. This term is $\langle v_x; v_x > 0\rangle/\delta q_x^*$. It originates from the mean velocity of the atoms crossing the dividing surface from A into B (see Chapter 14 for the definition of A and B) and the translational partition function of the same atom when it is in the "box" surrounding the region A*. The width δ of this region drops out. This is great, because we do not know the magnitude of δ.

The term

$$\exp\left[-\frac{V_T - V_A + \hbar(\omega_z^T + \omega_y^T - \omega_x - \omega_y - \omega_z)}{k_B T}\right] \qquad (15.25)$$

originates from the vibrational partition functions: it contains the energy of the transition state V_T, the lowest energy V_A of the compound A, and the *zero point energies* of all vibrations.

The rest of the formula

$$\frac{\left(1-\exp\left[-\frac{\hbar\omega_x}{k_BT}\right]\right)\left(1-\exp\left[-\frac{\hbar\omega_y}{k_BT}\right]\right)\left(1-\exp\left[-\frac{\hbar\omega_z}{k_BT}\right]\right)}{\left(1-\exp\left[-\frac{\hbar\omega_y^T}{k_BT}\right]\right)\left(1-\exp\left[-\frac{\hbar\omega_z^T}{k_BT}\right]\right)}$$

comes from the vibrational partition functions of A and A*. Each of these terms has a role to play in determining the properties of $k_{A \to B}$.

§11. *The Temperature Dependence of* $k_{A \to B}$. Empirically it has been found that the rate constant satisfies the Arrhenius formula (see §1 of Chapter 12):

$$k_{A \to B} = Pe^{-E_a/k_BT} \tag{15.26}$$

where the pre-exponential P and the activation energy E_a are independent of temperature.

At first sight the temperature dependence predicted by Eq. 15.24 has no resemblance to the Arrhenius formula, Eq. 15.26. Sometimes in science, as in personal relationships, first impressions can be misleading. The exponential term, Eq. 15.25, which is part of in Eq. 15.24 for $k_{A \to B}$, completely dominates the temperature dependence of $k_{A \to B}$. In spite of its complexity, Eq. 15.24 for $k_{A \to B}$ gives a temperature dependence very close to the Arrhenius formula. I will show in the next section that this is true.

A Numerical Study of $k_{A \to B}$

§12. *Introduction.* I have now all the information needed for calculating the rate constant as a function of temperature and the mass of the chemisorbed particle. From Eq. 15.8, $V_A = -1.6794$ eV. From Eq. 15.18, $V_T = -1.224$ eV.

The vibrational frequencies of the "reactant," ω_x, ω_y, ω_z, are given by Eq. 15.10. The vibrational frequencies around the transition state are given by Eqs 15.21 and 15.22. I can calculate the rate constant by using these data in Eq. 15.24.

I study first the temperature dependence and then I look at the *isotope effect.*

§13. *The Temperature Dependence of* $k_{A \to B}$ *is Consistent with the Arrhenius Formula.* The first question we address is whether the results calculated from Eq. 15.24 are compatible with the Arrhenius formula, Eq. 15.26. To test this,

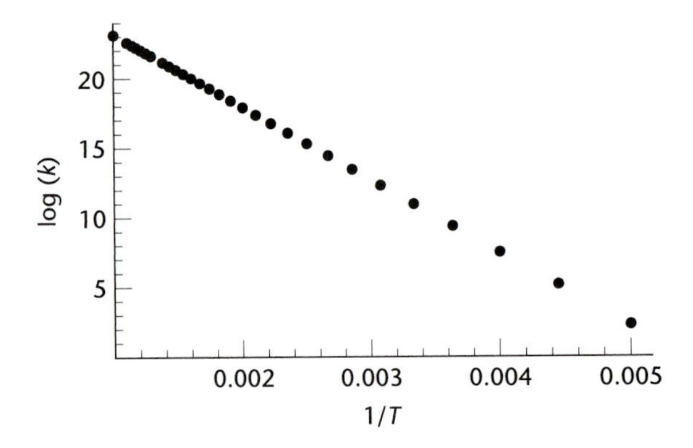

Figure 15.1 An Arrhenius plot of the rate constant k calculated from Eq. 15.24. T is in kelvin and k is in s^{-1}.

I calculate $k_{A \to B}$ for a variety of temperatures and plot $\ln k_{A \to B}(T)$ versus $1/T$. If the Arrhenius equation is correct the plot is a straight line, since

$$\ln k_{A \to B} = \ln P - \frac{E_a}{k_B T} \tag{15.27}$$

The slope is $-E_a/k_B$ and the intercept is $\ln P$. The plot is made in Cell 11 of Workbook SM15.1 and is reproduced in Fig. 15.1. As you can see, we obtain a straight line: our formula behaves just as Arrhenius guessed it would.

While graphs are nice, retrieving quantitative information from them is inaccurate. To determine the activation energy and the pre-exponential, I use least-squares fitting, to fit the rate constant $k_{A \to B}(T)$ given by Eq. 15.24 to the Arrhenius formula. For this I vary P and E_a to minimize the error

$$e(P, E_a) \equiv \sum_{i=1}^{n} (k_{A \to B}(T_i) - P \exp[-E_a/k_B T_i])^2 \tag{15.28}$$

Here $T_1, \ldots, T_N$ are N values of the temperature and $k_{A \to B}(T_i)$ is the rate constant calculated at the temperature T_i from Eq. 15.24. This procedure finds the values of E_a and P for which $P \exp[-E_a/k_B T]$ is closest to $k_{A \to B}(T)$. Least-squares fitting was discussed in detail in Chapter 3 of my book on thermodynamics.

The calculation (see Cell 12 of Workbook SM15.1) gives

$$E_a = 0.4548 \text{ eV} \tag{15.29}$$

and

$$P = 3.766 \times 10^{12} \text{ s}^{-1} \tag{15.30}$$

The energy of the transition state, $V_T = -1.224$ eV, minus the energy of the reactant $V_A = -1.6754$ eV is

$$V_T - V_A = 0.4513 \text{ eV} \tag{15.31}$$

This is very close to the activation energy E_a obtained by fitting $k_{A \to B}$ with the Arrhenius formula.

Why isn't the term containing the zero point energy (Eq. 15.25) making a large contribution to $k_{A \to B}$? In Cell 13 of Workbook SM15.1, I calculated the magnitude of this term. The results are reproduced in Table 15.1. As you can see, the contribution Eq. 15.25 to $k_{A \to B}$ (see Eq. 15.24) is unimportant because $a = \hbar[\omega^T + \omega_z^T - \omega_x - \omega_y - \omega_z] = 0.018$ eV (column 2 of Table 15.1) is much smaller than $V_T - V_A = 0.4513$ eV. Moreover, the magnitude of a is comparable

Table 15.1 The contributions of some of the terms in k_{A-B} to the temperature dependence. The zero-point energy contribution Zp is $\hbar(\omega_y^T + \omega_z^T - \omega_x - \omega_y - \omega_z)$.

T (K)	Zp (eV)	$k_B T$ (eV)	$\exp\left[-\frac{Z_p}{k_B T}\right]$	$\exp\left[-\frac{V_T - V_A}{k_B T}\right]$
200	1.86×10^{-2}	1.72×10^{-2}	0.68	4.23×10^{-12}
300	1.86×10^{-2}	2.58×10^{-2}	0.77	2.62×10^{-8}
400	1.86×10^{-2}	3.45×10^{-2}	0.82	2.06×10^{-6}
500	1.86×10^{-2}	4.31×10^{-2}	0.86	2.82×10^{-5}
600	1.86×10^{-2}	5.17×10^{-2}	0.88	1.62×10^{-4}
700	1.86×10^{-2}	6.03×10^{-2}	0.89	5.63×10^{-4}
800	1.86×10^{-2}	6.89×10^{-2}	0.91	1.43×10^{-3}
900	1.86×10^{-2}	7.76×10^{-2}	0.92	2.97×10^{-3}
1000	1.86×10^{-2}	8.62×10^{-2}	0.93	5.31×10^{-3}

to that of k_BT, which means that $\exp[-a/k_BT]$ is close to 1 at the temperatures on interest. Contrast this with the rapid change of $\exp[-V_T - V_A/k_BT]$ in column 5 of Table 15.1. Clearly, the latter term dominates the temperature dependence.

The calculations reported above were made for $m = 40$. If we were interested in hydrogen ($m = 1$), the vibrational frequencies would be larger and the zero point energy would contribute more to the rate constant.

Exercise 15.3

Calculate $k_{A \to B}$ for $m = 1$. Examine whether the zero point contribution is negligible in this case.

§14. *Is this General?* My experience has been that in most cases the Arrhenius formula works well and the activation energy is $V_T - V_A$. Exceptions are proton-, hydrogen-, and electron-transfer reactions.

In the case of electron-transfer quantum tunneling (see quantum mechanics) is fairly efficient and because of this the activation energy is lower than $V_T - V_A$. At very low temperatures, tunneling dominates and the Arrhenius formula no longer works; in many such cases the rate constant is temperature independent. However, at this temperature the rate is very small and it is often of no interest to practical chemists.

The theory presented here does not include tunneling. In the case of electron transfer reactions, tunneling is so important that the theory described here is unusable.

§15. *The Isotope Effect.* When you look at Eq. 15.24 for $k_{A \to B}$, you see that this quantity depends explicitly on the mass of the chemisorbed particle. Besides the dependence displayed explicitly, there is also an implicit dependence on mass through the vibrational frequencies ω_x, ω_y, ω_z, ω_y^T, ω_z^T. These quantities are all proportional to $1/\sqrt{m}$ (see Eqs 15.10, 15.20, and 15.21). Obviously, changing the mass will affect the hopping rate. Let us try to understand how the rate constant depends on mass and how to use this dependence in practical chemistry.

Consider two kind of measurements performed by experimentalists A and B. A heard about this mass dependence and decided to measure the mobility of H and He atoms on the surface. B thought that this was a stupid idea and made measurements on the mobility of H and D (D is deuterium).

It is true that in his experiments A changed the mass from 1 to 2. This is a change of 100% and it is likely to make a detectable difference in $k_{A \to B}$. If A was as stupid as B thinks he is, he would have compared thorium (mass 232.04) with palladium (mass 231.04). Here the relative change in the mass is so small that $k_{A \to B}$ is practically unaffected by it.

While choosing to work with H and He is better than choosing thorium and palladium, A did go wrong in one respect. H has one electron and He has two. Because of this, H tends to bind strongly to surfaces while He barely binds. The values of V_A, V_T, ω_x, ω_y, ω_z, ω_y^T, ω_z^T for H will be much larger than for He. The differences in these quantities are so large that they will completely overshadow the effect of mass change.

B was clever. Hydrogen and deuterium have the same nuclear charge and the same number of electrons (they are isotopes). Therefore, to an exceedingly good approximation they have the same potential energy surface. The values of V_A, V_T, k_{xx}, k_{yy}, k_{zz}, k_{xx}^T for these two atoms are the same. As far as the rate constant is concerned the two atoms differ through mass only. The experiment performed by B studies the mass effect and only the mass effect. In the one performed by A, the mass and everything else changes.

In Cell 14 of Workbook SM15.1, I calculated $k_{A \to B}$ for H and D. The values for V_A, V_T, ω_x, ω_y, ω_z, ω_y^T and ω_z^T are those derived from the potential energy used in all examples given here. The results are shown in Table 15.2. As you can see, the effect is large and it can be easily measured.

§16. *So What?* The change in the rate constant when you replace one atom in a reactant with one of its isotopes is called *the isotope effect*. We now know that changing H with D causes a large isotope effect. This is very useful because there are many reactions in which a hydrogen atom (or a proton) changes its position in a molecule or is transferred from one molecule to another. If, when studying a new reaction, you suspect that it involves moving hydrogen atoms, the isotope effect can be used to confirm your suspicion. To do this replace, with a deuterium atom, the hydrogen atom that you suspect is moving. Then perform the reaction with this deuterated compound. If you are right, the rate constant for the deuterated compound is smaller. If the rate constant does not change, your hypothesis was wrong.

§17. *Warning.* The reaction we studied here is very simple. In most practical cases the mass dependence is more complicated, but the principle remains: replace a moving hydrogen with a deuterium and the reaction slows down.

Table 15.2 The hopping rates k_H and k_D for H and D atoms chemisorbed on a surface.

Temperature, T (K)	k_H (SV^{-1})	k_P (SV^{-1})	K_D/K_H
200	290.73	126.92	0.44
300	1.05×10^6	5.80×10^5	0.55
400	6.65×10^7	4.07×10^7	0.61
500	8.23×10^8	5.29×10^8	0.64
600	4.45×10^9	2.94×10^9	0.66
700	1.49×10^{10}	1.00×10^{10}	0.67
800	3.72×10^{10}	2.53×10^{10}	0.68
900	7.57×10^{10}	5.19×10^{10}	0.69
1000	1.34×10^{11}	9.24×10^{10}	0.69

I mentioned that we have neglected here the effect of tunneling on the rate constant. For reactions involving hydrogen migration tunneling is important at low temperature. It so happens that the tunneling rate depends dramatically on mass. Exchanging H with D has a very large effect when tunneling is important. It seems that isotope effects should be studied under conditions that make tunneling important and the isotope effect is large. Unfortunately, at the temperatures when tunneling is important the rate constant tends to be very small, and measuring it can be difficult. The rate for H may be very different from the rate for D, but both may be too small to measure.

Summary

§18. *How General is this Theory?* I described rate theory by using a very simple example. After four chapters of explanation and calculations, you might protest and tell me that there is nothing simple about migration of atoms on a surface. Nevertheless, it is simpler than almost everything else. This has the advantage of allowing us to focus on the physics without being drawn into complicated mathematics.

The procedure developed here (the method of derivation) is general and it always leads to a formula of the form

$$k_{\mathrm{A}\to\mathrm{B}} = \frac{q^*}{q_{\mathrm{A}}} C \qquad (15.32)$$

The ratio of the partition function q^* of the activated complex to the partition function q_A of the reactant always appears in the formula for the rate constant. This ratio contains the factor

$$\exp\left[-\frac{V_T - V_A}{K_B T}\right] \tag{15.33}$$

and leads to an Arrhenius temperature dependence.

For our example (see Eq. 14.14)

$$C = \frac{\langle v_x; v_x > 0\rangle}{\delta} \tag{15.34}$$

This is not general. To understand why, you need to remember how we obtained it (see Chapter 14, §8 and §9). It is the rate of escaping from the transition complex region A* into the product region B. The derivation used the fact that the ridge between A and B is a line perpendicular to the x axis. Because of this, the dividing surface was a plane. In general, the ridge is a multidimensional, curved object, and the term C is more complicated and harder to calculate.

We have also made the harmonic approximation in calculating the partition function. This often works well, but it is an approximation. Nowadays, this is no longer necessary. The ratio of the partition functions can be calculated numerically by a Monte Carlo method.

The largest errors in rate-constant calculations are made because we are unable to accurately calculate the potential energy surface. This causes errors in $V_{TS} - V_A$, which in turn (through its presence in the factor $\exp[-V_{TS} - V_A/k_B T]$) causes larger errors in $k_{A\rightarrow B}$. The potential energy surface is calculated by quantum mechanics and we are making rapid progress in obtaining reliable potential-energy surfaces for molecules of practical interest (we do well for small molecules). By the time you approach retirement, it is likely that you will be able to obtain reliable rate constants from computations.

APPENDICES

Appendix 1. Values of Some Physical Constants

Adapted from D.A. McQuarrie, *Statistical Mechanics*, Harper & Row, New York, 1973, which cites B.N. Taylor, W.H. Parker, and D.N. Langenberg, *Rev. Mod. Phys.* **41**, 375, 1969.

Quantity	Symbol	Value
Avogadro's number	N_A	6.0222×10^{23}
Planck's constant	h	6.6262×10^{-27} erg s
	$\hbar$	1.0546×10^{-27} erg s
Boltzmann constant	k	1.3806×10^{-16} erg/mol K
Gas constant	R	8.3143×10^{7} erg/mol K
		1.9872 cal/mol K
Speed of light	c	2.9979×10^{10} cm/s
Proton charge	e	4.8032×10^{-10}esu
Electron mass	m_e	9.1096×10^{-28} g
Atomic mass unit	amu	1.6605×10^{-24} g
Bohr magneton	μ_B	9.2741×10^{-21} erg/gauss
Nuclear magneton	μ_N	5.0509×10^{-24} erg/gauss
Permitivity of vacuum	ϵ_0	8.854187×10^{-12} C^2/N^2 m^2

Appendix 2. Energy Conversion Factors

Source: D.A. McQuarrie, *Statistical Mechanics*, Harper & Row, New York, 1973.

	ergs	eV	cm^{-1}	K
1 erg	1	6.2420×10^{11}	5.0348×10^{15}	7.2441×10^{15}
1 eV	1.6021×10^{-12}	1	8.0660×10^{3}	1.1605×10^{4}
1 cm^{-1}	1.9862×10^{-16}	1.2398×10^{-4}	1	1.4388
1 K	1.3804×10^{-16}	8.6167×10^{-5}	6.9502×10^{-1}	1
1 kcal	4.1840×10^{10}	2.6116×10^{22}	2.1066×10^{26}	3.3009×10^{26}
1 kcal/mol	6.9446×10^{-14}	4.3348×10^{-2}	3.4964×10^{2}	5.0307×10^{2}
1 atomic unit	4.360×10^{-11}	27.21	2.195×10^{5}	3.158×10^{5}

	kcal	kcal/mol	atomic units
1 erg	2.3901×10^{-11}	1.4394×10^{13}	2.294×10^{10}
1 eV	3.8390×10^{-23}	2.3119×10^{1}	3.675×10^{-2}
1 cm^{-1}	4.7471×10^{-27}	2.8588×10^{-3}	4.556×10^{-6}
1 K	3.2993×10^{-27}	1.9869×10^{-3}	3.116×10^{-6}
1 kcal	1	6.0222×10^{23}	9.597×10^{20}
1 kcal/mol	1.6598×10^{-24}	1	1.594×10^{-3}
1 atomic unit	1.042×10^{-21}	6.275×10^{2}	1

FURTHER READING

Here is a list of books that you can read with profit if you decide to expand and refine your knowledge of statistical mechanics. I will give the smallest number of books that covers the material. There are two selection criteria: I like the book and you have adequate background to read it without pain.

1. D.A. McQuarrie, *Statistical Thermodynamics*, University Science Books, Harper & Row Publishers, New York, 1973 and T.L. Hill, *An Introduction to Statistical Thermodynamics*, Dover Publications Inc., New York, 1986.

These two excellent books cover most of the material that a chemist who does not specialize in statistical mechanics needs to know. The mathematical level is more advanced than in my book and the explanations are more concise, but you should be able to profit by reading these books.

2. D. Chandler, *Introduction to Modern Statistical Mechanics*, Oxford University Press, New York, 1987.

This book, written by an outstanding researcher, is a concise introduction to statistical mechanics. It complements what you have read here and often brings up interesting points of view and delightful insights.

3. E. Schrödinger, *Statistical Thermodynamics*, Dover Publications Inc., New York, 1989.

Once you start to understand statistical mechanics, read this book. It is excellently written and gives you a deeper understanding of the fundamentals of the subject. If you want to go deeper into the subject read one or all the books given below.

4. D.A. McQuarrie, *Statistical Mechanics,* Harper & Row Publishers, New York, 1976; and University Science Books, Mill Valley CA, 2000.

5. T.L. Hill, *Statistical Mechanics: Principles and Selected Applications,* Dover Publications, Inc., New York, 1987.

6. G.F. Mazenko, *Equilibrium Statistical Mechanics,* John Wiley & Sons, New York, 2000.

INDEX